Simplified Design of Linear Power Supplies

Simplified Design of Linear Power Supplies

John D. Lenk

Butterworth–Heinemann

Boston London Oxford Singapore Sydney Toronto Wellington

Cover photograph courtesy of Linear Technology Corporation.

Recognizing the importance of preserving what has been written, it is the policy of Butterworth-Heinemann to have the books it publishes printed on acid-free paper, and we exert our best efforts to that end.

Library of Congress Cataloging-in-Publication Data

Lenk, John D.
Simplified design of linear power supplies / by John D. Lenk.
p. cm.
Includes index.
ISBN 0-7506-9506-4
1. Electric power supplies to apparatus—Design and construction.
2. Electric circuits, Linear—Design and construction. 3. Power electronics. I. Title.
TK7868.P6L455 1994
621.381'044—dc20 94-16077
CIP

British Library Cataloguing-in-Publication Data

A catalogue record for this book is available from the British Library.

Butterworth–Heinemann
313 Washington Street
Newton, MA, 02158

10 9 8 7 6 5 4 3 2 1

Printed in the United States of America

Greetings from the Villa Buttercup!
To my wonderful wife, Irene:
Thank you for being by my side all these years!

To my lovely family, Karen, Tom, Brandon, Justin, and Michael, and to our Lambie and Suzzie: Be happy wherever you are!

To Frank Satlow, Karen Speerstra, Aliza Lamdan, John Fuller, the U.K. people, and the EDN people: A special thanks for making me an international best seller, again (this is book number 79)!

And to my special readers: May good fortune find your doorway, bringing you good health and happy things. Thank you for buying my books!

Abundance!

Contents

Preface

This book has something for everyone. No matter what your skill level in electronics, this book will show you how you can immediately experiment with, test, interconnect, and design linear power supplies.

For experimenters, students, and serious hobbyists, the book provides sufficient information to design and build linear power supplies "from scratch." The design approach here is the same one used in all of the author's best-selling books on simplified and practical design.

The first six chapters provide the basics for all phases of practical design, including testing and interconnecting supplies. The final chapter includes well over 100 worked-out design examples, using the techniques described in the first six chapters.

Throughout the book, design problems start with approximations or guidelines for selecting all components on a trial-value basis, assuming a specific design goal and set of conditions. Then, using these approximate values in experimental circuits, the desired results (input/output voltage and current, line and load regulation, ripple rejection, noise, etc.) are produced by varying the test component values.

Although all circuit operations are described thoroughly where needed, mathematical theory is kept to a minimum. No previous design experience is required to use the design data and techniques described in this book. The reader need not memorize elaborate theories, or understand abstract math, to use the design data. With any linear supply it is possible to apply certain guidelines for selecting component values. These rules can be stated in basic equations, whose solutions require only simple arithmetic. Thus the book is ideal for the practical experimenter.

For service technicians and field-service engineers, an entire chapter is devoted to basic and advanced tests for linear power supplies. Three groups of tests are covered: The first, generally sufficient for most practical applications, can be performed with simple test equipment. The second describes advanced test techniques that apply to linear supplies and supply components designed for commercial use.

The chapter concludes with notes on external connections for linear supplies, including remote sensing.

If you are a working engineer responsible for designing and/or selecting linear power supplies, the variety of circuits and configurations described here should generally simplify your task. Not only does the book describe basic linear-supply designs, but it also covers the most popular forms of supply regulation, including zener, feedback, and op-amp. Integrated-circuit (IC) regulators and heat sinks are included, as are practical mounting and interconnection techniques for linear-supply components.

Chapter 1 is devoted to basic power-supply and linear-regulator circuits. These include half-wave, full-wave, bridge, voltage-doubling, and voltage-tripling circuits, as well as zener and adjustable-shunt (or programmable zener) regulation circuits.

Chapter 2 covers heat sinks and other temperature-related design problems for linear power supplies.

Chapter 3 discusses basic feedback-regulator circuits. These include series and shunt constant-voltage (CV) regulators, high-current and constant-current (CC) regulators, as well as overload protection for linear power supplies.

Chapter 4 is devoted to basic IC linear regulators. These include IC voltage and current regulators, as well as short-circuit and overload protection, current boosting, current foldback, shutdown, and three-terminal regulators.

Chapter 5 focuses on IC op-amps used as linear voltage regulators.

Chapter 6 presents test procedures and connections for linear power supplies. The procedures can be applied to a just-completed linear supply during design or experimentation, or to a suspect supply as part of troubleshooting.

Chapter 7 is devoted to design examples for linear power supplies, using off-the-shelf components. All of the design techniques discussed in the first six chapters are used as needed. These circuits can be put to immediate use as is or, by altering the component values, used as a basis for simplified design of similar linear supplies.

Acknowledgments

Many professionals have contributed to this book. I gratefully acknowledge their tremendous effort in making this work so comprehensive: it is an impossible job for one person. I thank all who contributed, directly or indirectly.

I give special thanks to Syd Coppersmith of Dallas Semiconductor, Rosie Hinojosa and Kellie Garcia of EXAR Corporation, Jeff Salter of GEC Plessey, Linda daCosta of Harris Semiconductor, David Fullager of Maxim, Linda Capcara of Motorola, Inc., Andrew Jenkins of National Semiconductor Corporation, Antonio Ortiz of Optical Electronics Inc., Lawrence Fogel of Philips Semiconductors, Lorraine Jenkins of Raytheon Company Semiconductor Division, Anthony Armstrong of Semtech Corpus Christi, Ed Oxner and Robert Decker of Siliconix Inc., Amy Sullivan of Texas Instruments, and Fred Swymer of Unitrode (Microsemi) Corporation.

I also thank Joseph A. Labok of Los Angeles Valley College for help and encouragement throughout the years.

Very special thanks to Frank Satlow, Karen Speerstra, Laurel DeWolf, Aliza Lamdan, John Fuller, the U.K. people, and the EDN people of Butterworth–Heinemann for having so much confidence in me. I recognize that all books are a team effort and am thankful that I am working with the New First Team!

And to Irene, my wife and super agent, I extend my thanks. Without her help, this book could not have been written.

CHAPTER 1

Linear Power Supply Basics

This chapter is devoted to basic power-supply and linear-regulator circuits. These include half-wave, full-wave, bridge, voltage-doubling, and voltage-tripling circuits, as well as zener and adjustable-shunt (or programmable-zener) regulation circuits.

1.1 Half-Wave Circuits

Figure 1–1 shows the basic half-wave power-supply circuit. The cathode of diode CR_1 is connected to filter capacitor C_1 and filter choke L_1. Another filter capacitor, C_2, is also used to smooth out the pulsating direct current. Resistance R_1 is placed across the supply output. (R_1 is not used in all circuits; instead, the output is connected directly to the load circuit.) Resistor R_1, if used, is known as a bleeder resistor, because it places a small current drain on the supply and helps stabilize the output.

If resistance R_1 is made up of a single tapped resistor or a series of several resistors, it is possible to take several different voltages from the bleeder network. For example, if the power-supply output is 30 V and the bleeder is made up of six equal-value resistors, the available voltages are 5, 10, 15, 20, 25, and 30 V, as shown.

The circuit output is a direct current but not necessarily a constant voltage. In practical applications, there is some variation in amplitude (known as ripple). The output voltage decreases slightly between cycles (negative peak of the ripple voltage) and then increases at the peak of each half cycle (positive peak of the ripple voltage).

Typically the peak-to-peak ripple voltage is expressed as a percentage of the total power-supply output voltage. For example, if the power supply produces 100 V of direct current across the bleeder resistor and the ripple is 3 V peak-to-peak, there is a 3% ripple (which would be totally unacceptable for almost any type of linear supply!). The amount of ripple for any given type of basic power-supply circuit (half-wave, full-wave, etc.) depends on the type of filter.

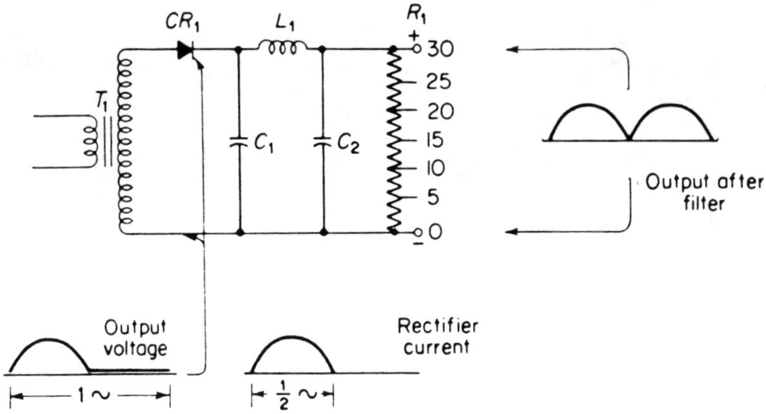

Figure 1–1. Basic half-wave power supply

1.2 Full-Wave Circuits

Figure 1–2 shows the basic full-wave power-supply circuit. Both the positive and negative alternations of the AC cycle are used in a full-wave supply. For this reason, the full-wave circuit requires two diodes and a transformer with a center tap. One lead of the transformer secondary is connected to diode CR_1; the opposite end of the secondary is connected to diode CR_2. The center tap is connected to the common or ground circuit. The total voltage across the transformer secondary is about twice the voltage that appears at the supply output.

The filter capacitors charge at a rate twice that for half-wave rectification, because current flows through the bleeder resistor in the same direction on both alter-

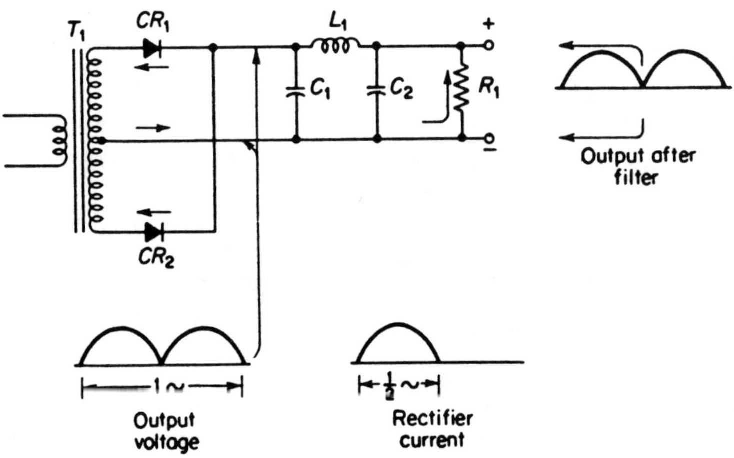

Figure 1–2. Basic full-wave power supply

nations or half cycles. Consequently, the ripple frequency is twice that of the half-wave-rectification ripple frequency.

Because the discharge time between peaks of full-wave pulsating direct current is only half that found in half-wave rectification (there is a shorter time between peaks), the filter capacitors have less time to discharge. This makes it easier to maintain a relatively high charge and makes the output smoother than that of a half-wave supply.

1.3 Bridge Circuits

Figure 1–3 shows the basic full-wave bridge power-supply circuit. The bridge circuit makes it possible to have full-wave rectification using a transformer without a center tap. Four rectifiers are used in the basic bridge circuit. (Also, there are rectifier packages that contain four rectifiers connected in the bridge configuration.)

When a positive alternation occurs, with the top end of the transformer secondary positive, current flows from the secondary bottom, through CR_3, the load or bleeder resistor R_1, and diode CR_2 to the secondary top. On the next alternation, when the bottom end of the transformer secondary is positive, current flows from the secondary top, through CR_1, R_1, and CR_4 to the secondary bottom.

Full-wave rectification occurs since current flows through the bleeder or load resistance in the same direction on both half cycles. This produces both a higher ripple frequency and more efficient filtering.

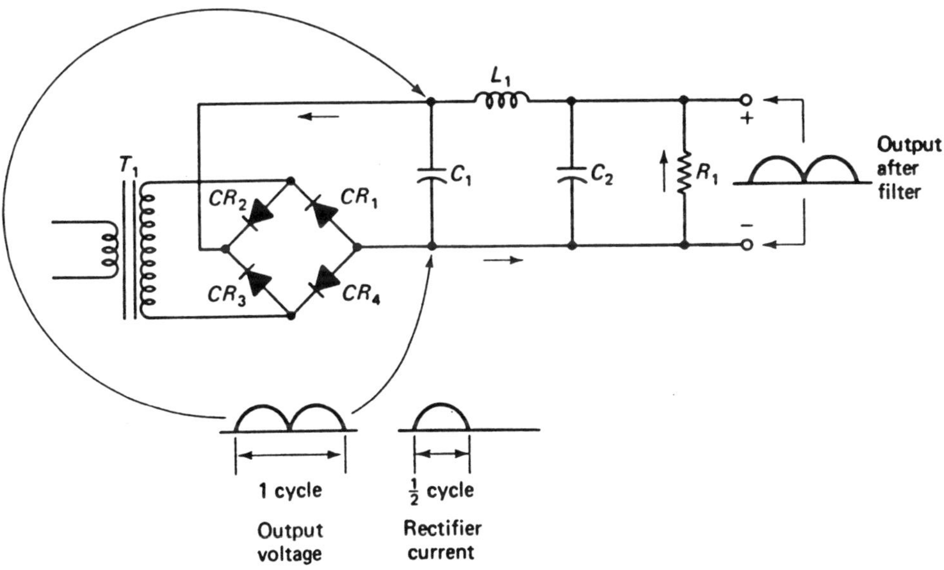

Figure 1–3. Basic full-wave-bridge power supply

1.4 Basic Power-Supply Design

Figure 1–4 shows the design schematic for a basic power supply. Notice that the full-wave bridge with only one filter capacitor is selected. This is generally the most efficient circuit for linear-regulator applications. For two reasons, elaborate filters are not needed in present-day equipment. First, the added weight and bulk produced by large filter chokes cannot be tolerated in solid-state/integrated-circuit (IC) equipment (unlike ancient vacuum-tube equipment). Second, and more important, a regulator of some type (zener, transistor, IC, three-terminal, etc.) follows the basic power-supply circuit (transformer, diodes, capacitor). Such regulators filter and control the power-supply output much better than any type of filter.

The following paragraphs describe the simplified design approach for this circuit. Before actually assembling and testing the circuit, read the heat-sink design considerations in Chapter 2 and the test procedures in Chapter 6.

Notice that the following design procedures assume that it is necessary to make up the bridge with four separate diodes. As a practical matter, if you can find a single bridge-rectifier package with the required voltage and current ratings, use the package instead of individual diodes.

1.4.1 Power Transformer

In present-day equipment, T_1 is usually a step-down transformer. Typically, the primary voltage is 110–120 V (or 220–240V) and the secondary is between 12 and 50 V. If the transformer is used with a full-wave bridge rectifier and a single-capacitor filter, the DC output of the supply is about 1.3 times the transformer sec-

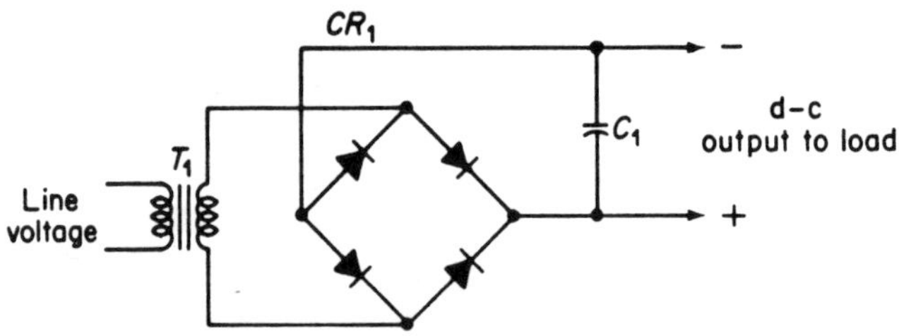

d-c output ≈ 1.3 × secondary voltage
Transformer power (or VA) rating > 1.3 × d-c output power
Load resistance = output voltage/output current
Power supply resistance = (no-load V − full load V)/current
C_1 working voltage > 1.3 × output voltage
C_1 capacitance (in μF) ≈ 200,000/load resistance × max ripple
Secondary voltage ≈ required d-c output/1.3
d-c power output = output voltage × load current
CR_1 ratings = see text

Figure 1–4. Design schematic for basic power supply

ondary voltage (e.g., a T_1 secondary of 10 V produces about 13 V DC output).

In addition to voltage, transformers are rated as to power capability—usually a volt-ampere (VA) rating rather than a true AC-power rating. As a rule, the transformer power rating should be at least 1.3 times the DC-output power of the complete circuit (e.g., if 10 W of DC power is required, the transformer should be capable of at least 13 W).

1.4.2 Load Resistance and Power-Supply Resistance

These two terms are often confused. Neither value can be measured directly but must be calculated on the basis of voltage and current.

Load resistance (or effective load impedance) is found when the output voltage across the load is divided by the load current. For example, with a DC output of 10 V and a load current of 2 A, the load impedance is 5 ohms.

Power-supply resistance (or internal impedance) is found when the change in output voltage is divided by a change in load current (see Chapter 6). In addition to this basic method, it is also possible to calculate the internal resistance using a small load and a full load (instead of the full load and no load described in Chapter 6). The difference in voltage is divided by the difference in load current. For example, if the small-load voltage is 28 V with 0.5 A and the full-load output is 25 V with 2 A, the power-supply resistance is (28 – 25)/(2 – 0.4) = 3/1.5 = 2 ohms.

1.4.3 Diode Characteristics

There are four basic characteristics that must be considered for any type of diode: maximum reverse voltage, forward voltage, reverse current, and forward current.

Maximum reverse voltage (also known as peak inverse voltage, or PIV) is the amount of reverse voltage that a diode can withstand without breakdown. Data sheets often list two values for reverse voltage: average (or normal) and peak (or maximum) reverse voltage. As a guideline for full-wave bridge rectifiers, the average-reverse-voltage rating should be about twice the DC-output voltage. In any event, the reverse-voltage rating must be greater than either the secondary-output voltage or the DC-output voltage.

Forward voltage (or forward-voltage drop) is the amount of drop across the diode in the forward-bias condition (diode conducting). Ideally, the forward-voltage drop should be zero. Typically, the drop is about 0.5 to 1.0 V. (Notice that the diodes drop the secondary-output voltage, but this is offset by the voltage buildup across the capacitor.)

Reverse current (or leakage current) is the amount of current flow when the diode is reverse-biased (diode not conducting). Ideally, reverse current should be zero. Typically, reverse current is less than a few microamperes (at most a few milliamperes).

Forward current is the maximum current capacity of the diode in the forward-bias condition. As a guideline for full-wave bridge rectifiers, the forward-current rating of each diode must be greater than the DC-output current and should be about twice the DC-output current.

1.4.4 Capacitor Characteristics

Capacitors used in full-wave bridge supplies are electrolytics, because a high capacitance value is required. The voltage rating of the capacitor should be at least 1.3 times the DC-output voltage. The capacitance rating is selected on the basis of allowable ripple and load resistance, as shown by the equation in Fig. 1–4. For example, assuming a load resistance of 50 ohms, and a maximum permissible ripple of 5%, the capacitance values should be 200,000/(50 × 5), or 800 μF.

1.4.5 Design Example: Basic Power Supply

Assume that the circuit in Fig. 1–4 is to provide an output of 28 V at 400 mA (or 0.4 A), with a ripple of not more than 3%. A line supply of 110–120 V is available.

With a required 28 V output, the T_1 secondary should be 28/1.3 V, or 21.5 V. With an output of 11.2 W (28 V × 0.4 A), the volt-ampere (or power) rating of T_1 should be 11.3 × 1.3, or 14.56 W, rounded off to 15 W. The load resistance is 28/0.4, or 70 ohms.

The diode characteristics should be as follows:

- Maximum reverse voltage is 28 × 2, or 56 V. In simplified design, any reverse-voltage rating above about 30 V should be satisfactory.
- Forward-voltage drop should not exceed 1 V.
- Reverse current should not exceed 100 μA.
- Forward current is 0.4 × 2, or 0.8 A (800 mA).

Notice that these characteristics are for individual diodes. As discussed, it is possible to use a bridge-rectifier package instead of the four diodes. In that case, the data-sheet ratings for the package must be considered as the ratings for individual diodes.

The capacitor C_1 should have a voltage rating of 28 × 1.3, or 36.4 V. Any working voltage of about 36 V should be satisfactory. With a load resistance of 70 ohms and a maximum ripple of 3%, the capacitance value of C_1 should be 200,000/(70 × 3), or 952.3 μF (any value between 900 and 1000 μF should be satisfactory, with 1000 μF providing the best ripple rejection).

1.5 Voltage-Doubling Circuits

Figure 1–5 shows the basic voltage-doubling circuit. Using such circuits, it is possible to increase an available AC voltage without a transformer. As shown, the basic voltage-doubling circuit requires two rectifier diodes and two capacitors.

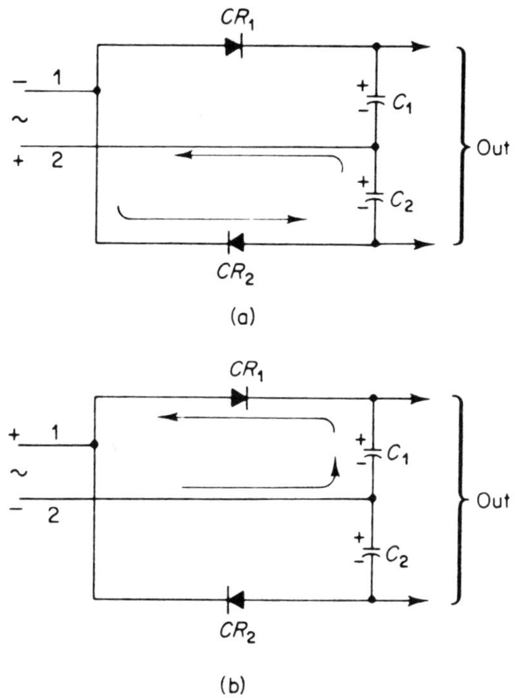

Figure 1–5. Basic voltage-doubling circuit

When the AC-line alternation is such that terminal 1 is negative and terminal 2 is positive (as in Fig. 1–5(a)), current flows from terminal 1 in the direction of the arrows. The current path from terminal 1 includes CR_2 and C_2 before a return to positive terminal 2. Capacitor C_2 is charged during this current flow with a polarity as shown.

During the next alternation, when terminal 1 is positive and terminal 2 is negative, current flows as shown in Fig. 1–5(b). As a result, current flows from terminal 2 toward C_1, and charges capacitor C_1 to the peak of the AC voltage. Current continues through diode CR_1 to the positive terminal 1.

Capacitors C_1 and C_2 are charged to the peak values of the AC-line voltage on alternate half cycles. The DC-output voltage is about double that of the AC voltage. The two capacitors are, in effect, in series (and the polarities add to increase the total voltage).

1.6 Voltage-Tripling Circuits

Figure 1–6 shows the basic voltage-tripling circuit. With such circuits it is possible to triple an available AC voltage without a transformer. As shown, the basic voltage-tripling circuit requires three rectifier diodes and three capacitors.

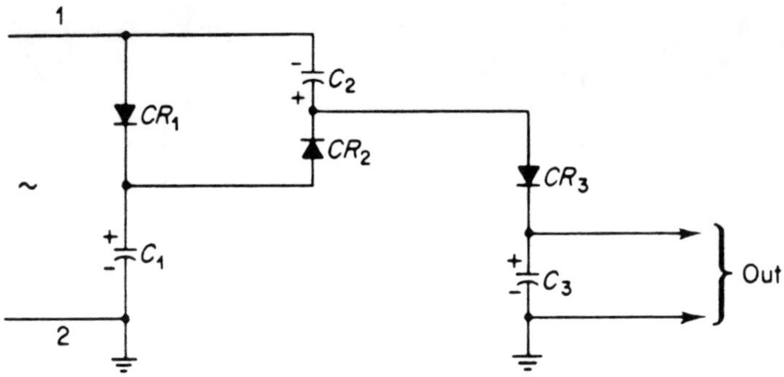

Figure 1–6. Basic voltage-tripling circuit

When the AC-line alternation is such that terminal 1 is positive and terminal 2 is negative, current flows from terminal 2 through CR_1 to terminal 1. Diode CR_1 is forward-biased under these conditions, permitting capacitor C_1 to be charged to the AC-line voltage.

During the next alternation, when terminal 1 is negative and terminal 2 is positive, CR_1 is reverse-biased and does not conduct. Instead, CR_2 is forward-biased and permits current flow toward terminal 2 through C_1. The charge that is already across C_1 adds to the AC-line voltage and places a charge across C_2. This charge is proportional to the line voltage added to the charge already across C_1.

During the next cycle, when terminal 1 is again positive and terminal 2 is negative, C_1 is again charged as before. At the same time, CR_3 conducts and charges C_3. When CR_3 conducts, the charge across C_3 is composed of the line voltage plus the existing charge across C_2. Because the C_2 charge is already double voltage, this is added to the line voltage and produces a triple voltage across C_3.

1.7 Basic Regulator Circuits

The amount of voltage that can be taken from a basic power-supply circuit is determined by the voltage available at the source (line or transformer), less any voltage drop through the rectifiers and filter. This voltage drop is determined by the amount of current flowing through the power-supply components (rectifiers, etc.). As more current is drawn from the power supply, the voltage drop across the components increases and the output voltage is reduced. When less current is drawn from the supply, the output is increased because the components have a minimum voltage drop.

The variation of voltage output with respect to the amount of current drawn from the supply is known as *voltage regulation* or simply *regulation.* Such regulation is expressed as a percentage, with the value being determined as follows:

$$\%\ \text{regulation} = \frac{(\text{no-load voltage}) - (\text{full-load voltage})}{\text{full-load voltage}} \times 100.$$

This equation applies to voltage regulation and takes into account the proportions of voltage increase and decrease with a change of load on a power supply. The smaller the difference between full-load and no-load voltages, the better the regulation.

1.7.1 Filter-Capacitor Regulation

Filter capacitors (such as C_1 in Fig. 1–4) provide some measure of regulation for the basic power-supply circuit. Should the output voltage drop, the capacitors discharge and serve to keep the voltage constant. The regulation provided by the filter capacitor is sufficient for a few applications. However, in most cases, the power-supply output must be maintained at some critical voltage or current. This requires some form of linear or switching regulator.

1.7.2 Linear and Switching Regulators

A number of solid-state circuits and devices have been developed to provide both voltage and current regulation. There are two basic approaches to such regulators: linear and switching. This book concentrates on the linear regulator, both discrete-component and IC, as well as on combination discrete/IC circuits.

The most common method for linear voltage regulation involves zener diodes. Zeners maintain the voltage drop across their terminals at a constant value, no matter what current is being drawn (within limits). The basic zener is also used as a voltage reference in other linear-regulator configurations.

1.8 Basic Zener Regulator Design

Figure 1–7 shows the design schematic for a zener regulator. In the simplest form, a zener regulator consists of a series resistance R_S and a shunt-connected zener CR_1. The value of R_S is set by the load requirements. If R_S is too large, the zener is not able to regulate at large values of load current. If R_S is too small, the zener dissipation rating (power or wattage rating) may be exceeded at low-load current values. Although the zener provides the simplest method for voltage regulation, certain limits must be considered in design.

The following paragraphs describe the simplified design approach for this circuit. Before actually assembling and testing the circuit, read the heat-sink design considerations in Chapter 2 and the test procedures in Chapter 6.

1.8.1 Zener Characteristics

Data sheets do not always show the same characteristics for zeners. In simplified design, the critical characteristics are zener (or avalanche) voltage, maximum current, and power capability.

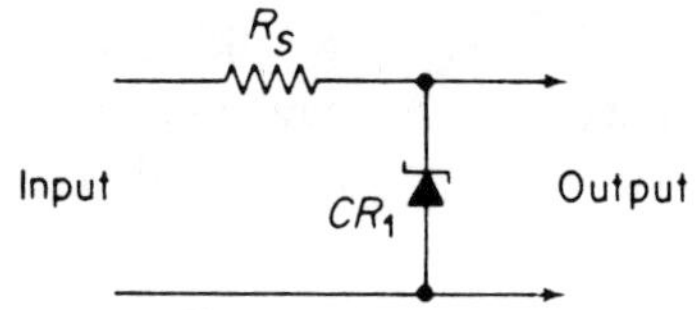

Output voltage = Zener voltage

Zener power dissipation = Zener voltage × Zener current

Minimum input voltage > 1.4 × Zener voltage

Zener voltage ≈ 0.7 × lowest input voltage

Safe Zener power ≈ 3 × load power

$$R_S \text{ (in ohms)} \approx \frac{(\text{max input voltage} - \text{Zener voltage})^2}{\text{safe power dissipation}}$$

Input voltage variation < 30 % of max input voltage

Load current = Zener voltage/load resistance

Load power = Zener voltage × load current

Safe Zener current ≈ 3 × load current

Figure 1–7. Design schematic for zener regulator

The power-dissipation capability of a zener can be considered as the zener voltage multiplied by the maximum current rating. Zeners are rarely rated for greater than a few watts of power dissipation. This rating is a room-temperature (typically 25°C) power dissipation and must be derated for higher temperature.

In theory, the zener maintains the output at the avalanche voltage (with a tolerance from about 0.1 to 0.9 V) despite any change in input and load. In practice, if the input voltage falls below about 1.4 times the zener voltage, the zener does not go into the avalanche condition and there is no regulation. On the other hand, if the input voltage rises to a point where the maximum current is exceeded, the zener overheats (and possibly is destroyed). Current through the zener is limited by series resistor R_S.

1.8.2 Zener Trial Values

Figure 1–7 shows the equations for trial values in zener-regulator circuits. The following summarizes the use of these equations. The zener voltage should be about 0.7 times the lowest input voltage. For example, if the input voltage varies between 10 and 15 V, the zener voltage should be 10 × 0.7, or 7 V.

If the design problem is stated in reverse, the minimum input voltage must be 1.4 times the desired zener voltage. For example, if the required zener (or regulated output) voltage is 10 V, the minimum input voltage from the basic power supply is 10 × 1.4, or 14 V.

A safe power-dissipation rating for the zener is three times the load power. With a 250-mW load (say 10 V across 400 ohms), a zener capable of dissipating 750 mW (or more) should be satisfactory.

The value of R_S is found using the maximum input voltage, the desired zener voltage, and the safe power dissipation. Using the same 10- to 15-V input, a desired zener voltage of 7 V, and a safe power dissipation of 600 mW, the first trial value of R_S is $(15 - 7)^2/0.6$, or 106 ohms (use the next-highest standard, 110 ohms). Notice that if the input voltage varies by more than about 30% (of the highest input voltage), it is difficult if not impossible to provide regulation with a zener.

Although the circuit of Fig. 1–7 can be assembled in experimental form using the trial values just calculated, it is convenient to check the design on paper, using maximum and minimum input-voltage values. If the trial value for R_S has been properly calculated, the maximum input voltage produces a current that can safely be dissipated by the zener, while the minimum input voltage provides more than enough current to meet the load requirements. The on-paper calculations for maximum and minimum input voltage are included in the following design example.

1.8.3 Design Example: Basic Zener Regulator

Assume that the circuit of Fig. 1–7 is to provide a regulated 7.5-V output to a 250-ohm load, that the maximum input voltage is 15 V, and that 7.5-V zeners are available. Find the minimum input voltage that is practical, the current (or power) rating for the zener, and the optimum value for series resistor R_S.

With a 7.5-V zener, the minimum input voltage is 7.5×1.4, or 10.5 V. With a 7.5-V output across a 250-ohm load, the load current is 7.5/250, or 30 mA. The load power is 7.5×0.03, or 225 mW.

A safe power-dissipation rating for the zener is 225×3, or 675 mW. The nearest standard value is (probably) a 1-W zener. With a safe power-dissipation rating of 675 mW, a zener voltage of 7.5, and a maximum input voltage of 15 V, the first trial value for R_S is $(15 - 7.5)^2/0.675$, or 83.3 ohms. The nearest 10% standard value is 82 ohms. With a load current of 30 mA, the current rating of the zener is 30×3, or 90 mA. A 100-mA rating provides an added measure of protection.

The design values (82 ohms for R_S and a zener of 7.5 V with a 100-mA rating) should be checked against the maximum and minimum input voltages, as follows. The maximum input current (combined zener current and load current) is equal to the drop across R_S ($15 - 7.5 = 7.5$ V) divided by the resistance (82), or 91 mA. With a normal operating load current of 30 mA, the zener normally passes only 61 mA ($91 - 30 = 61$ mA). If the load current drops to zero, the full 91 mA flows through the zener, so the 100-mA rating is adequate.

Maximum power to the zener is 7.5×0.09, or 675 mW, as calculated. A 1-W rating is adequate. Maximum power dissipated by R_S is also 675 mW (7.5-V drop, 0.09 A). So R_S should have a 1-W dissipation rating.

As a final calculation before going to the experimental stage, check that the minimum input voltage provides more than enough current to meet the load requirements. Minimum input voltage was calculated as 10.5 V. Using this value and a 7.5-V zener, the drop across R_S is $10.5 - 7.5$, or 3 V.

With a 3-V drop across 82 ohms, the minimum current is 3/82, or 36 mA. Under normal conditions, the load draws 30 mA, with the zener drawing the remaining 6 mA. If the current is less than 30 mA, the value of R_S must be decreased.

1.9 Extended Zener Voltage Range

Sometimes it is necessary to regulate a voltage not available in standard zener diodes. This problem can be overcome by various circuit arrangements. For example, it may be necessary to regulate a higher voltage with high current, beyond the capability of existing zeners. (Generally, high-voltage zeners have low current ratings.) In that case it is possible to connect several zeners in series for a given voltage.

Figure 1–8 shows how several zeners are connected in series, where the total regulated voltage is the sum of the individual zener voltages. If the zener voltages are as shown in Fig. 1–8 (5, 10, and 15 V), the total regulated output is 30 V.

1.9.1 Extended Zener Design Considerations

The design considerations for the circuit of Fig. 1–8 are similar to those of Fig. 1–7. The zeners need not have equal breakdown voltages because the arrangement is self-equalizing. However, the handling ability of each zener should be the same. Also, the current ranges should be similar or the loads so arranged to avoid damaging any of the zeners.

In practice, if the input voltage falls below about 1.4 times the total zener voltage of the series string, some or all of the zeners do not go into avalanche condition,

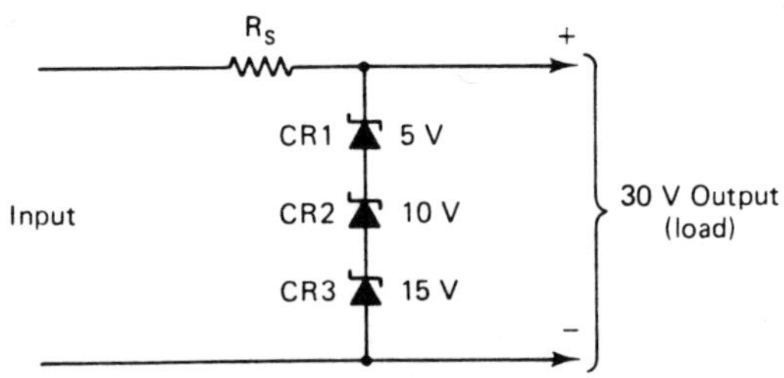

Output voltage = total series zener voltage
Individual zener power dissipation = individual zener voltage X zener current
Minimum input voltage > 1.4 X total series zener voltage
Total series zener voltage ≈ 0.7 X minimum input voltage
Safe power for each zener ≈ 0.3 X individual zener power dissipation
Total load current = total series zener voltage/load resistance
Total load power = total series zener voltage/total load current
Safe zener current ≈ 3 X total load current

$$R_S \text{ (in ohms)} \approx \frac{(\text{max input voltage} - \text{total series zener voltage})^2}{\text{safe power dissipation}}$$

Figure 1–8. Design schematic for zeners where output is sum of individual zeners

and there is no regulation. On the other hand, if the input voltage rises to a point where the maximum current of any one zener is exceeded, the zener overheats (and could be destroyed).

1.9.2 Extended Zener Trial Values

The total zener voltage should be about 0.7 times the lowest input voltage. In the reverse, the minimum input voltage must be 1.4 times the total zener voltage. If each of the three zeners is 7 V, the total zener voltage is 21 V, and the minimum input voltage is $21 \times 1.4 = 29.4$ V.

A safe power-dissipation rating for each zener is 3 times the load power. With a 100-mW load, each of the zeners should be capable of dissipating 300 mW. The value of R_S is found using the safe power dissipation, the total series zener voltage, and maximum input voltage. As in the case of the basic zener circuit, if the input voltage varies by more than about 30% of the highest input voltage, it is difficult to design a practical regulation circuit (with one or more zeners).

Before going to the experimental stage, check the design on paper. With a correct value for R_S, the maximum input voltage produces a current that can safely be dissipated by the zener with the lowest current rating, while the minimum input voltage provides more than enough current to meet the load requirements.

1.9.3 Design Example: Extended Zener

Assume that the circuit of Fig. 1–8 is to provide a regulated 22.5-V output to a 75-ohm load, that the highest current rating of an available 22.5-V zener is 100 mA, that 7.5-V zeners are available, and that the maximum input voltage is 40 V. Find the minimum input voltage that is practical, the current (or power) rating for each zener, and the optimum value for series resistor R_S.

With a required 22.5 V (to be provided by three 7.5-V zeners), the minimum input voltage is 22.5×1.4, or 31.5 V. With a 22.5-V output across a 75-ohm load, the load current is 22.5/75, or 330 mA. This exceeds the 100-mA capability of the available 22.5-V zener, so three 7.5-V zeners must be used. The total load power is 22.5×0.3, or 6.75 W. The power dissipated by each of the three zeners is 7.5×0.3, or 2.25 W.

A safe power-dissipation rating for each zener is 3 times the individual power, or 6.75 W. However, the nearest standard value is probably a 10-W zener. With a load current of 330 mA, the current rating of each zener is 300×3, or 900 mA.

With a safe power-dissipation rating of 6.75 W, a total zener voltage of 22.5 W, and a maximum input voltage of 40 V, the first trial value for R_S is $(40 - 22.5)^2/6.75$, or 45 ohms. The nearest 10% standard is 47 ohms.

The design values (47 ohms for R_S and three zeners of 7.5 V with 900-mA ratings) should be checked against the maximum and minimum input voltages as follows. The maximum input current (combined zener current and load current) is equal to the drop across R_S (40 – 22.5, or 17.5 V) divided by the resistance (47 ohms), or 372 mA. This provides the full 300 mA for the load, plus 72 mA through the zeners.

However, with a minimum input voltage calculated at 31.5 V and a zener of 22.5 V, the drop across R_S is 9 V. With a 9-V drop and a 47-ohm resistance, the current is slightly less than 220 mA, *not enough to supply the 300-mA load.* The 47-ohm value for R_S is too high. In that event, cut the calculated value in half to 23.5 ohms, or to the nearest standard of 27 ohms.

With a minimum drop of 9 V across 27 ohms, the current is 333 mA, or more than enough to meet the maximum load of 200 mA. Going back to the maximum input voltage of 40 V, the input current is 40 – 22.5, or 17.5 V; and 17.5 V/27 = 650 mA. The 900-mA rating for each zener is adequate.

1.9.4 Using a Zener as a Series Element

As shown in Fig. 1–9, a zener can be used as a series element. This circuit is generally used only where a small voltage drop is required. (The circuit of Fig. 1–8 is used for large voltage drops.)

In the series circuit of Fig. 1–9, the drop across the zener is 5.6 V, lowering the 28-V input to 22.4 V. Note that the entire load current, plus the current through R_S, passes the series zener. *The total current must be used* when calculating the required power-handling availability (wattage rating) of the zener.

1.9.5 Using Zeners as a Voltage Divider

Several zeners can be connected in series to act as a divider to obtain several regulated voltages, as shown in Fig. 1–10. Using the four zeners shown, it is possible to obtain the following 10 voltages across the corresponding terminals: 3.9 DE, 6.8 CD, 10.7 CE, 12 BC, 18.8 BD, 22.7 BE, 27 AB, 39 AC, 45.8 AD, and 49.7 AE.

1.9.6 Using Zeners for Lower Output Voltages

It is possible to arrange zeners to produce a regulated output at a voltage lower than that of the zener. Such a circuit is shown in Fig. 1–11, where two zeners are used. The output is the regulated difference voltage (8.2 – 6.8 = 1.4 V). This

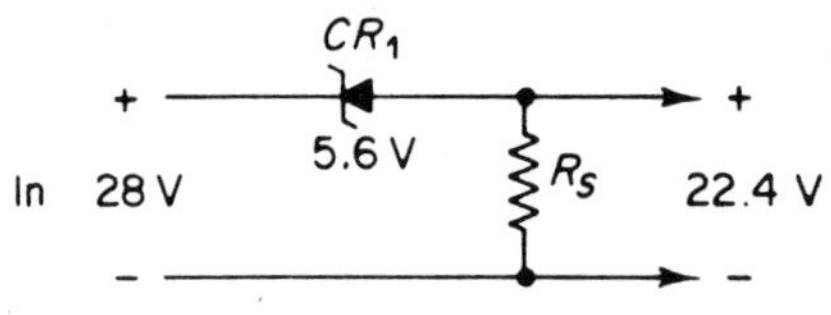

Figure 1–9. Using a zener as a series element

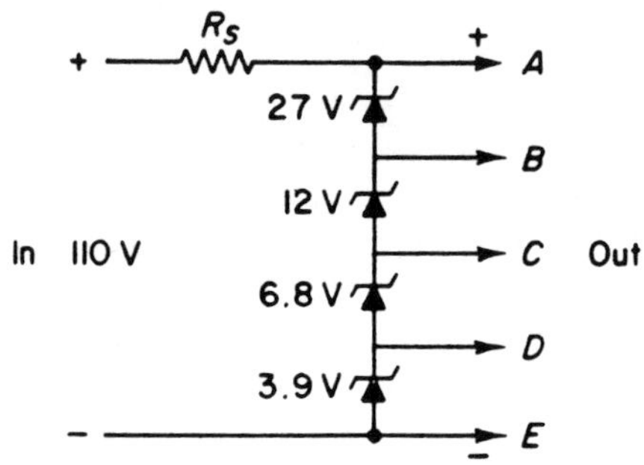

Figure 1–10. Using zeners as voltage divider

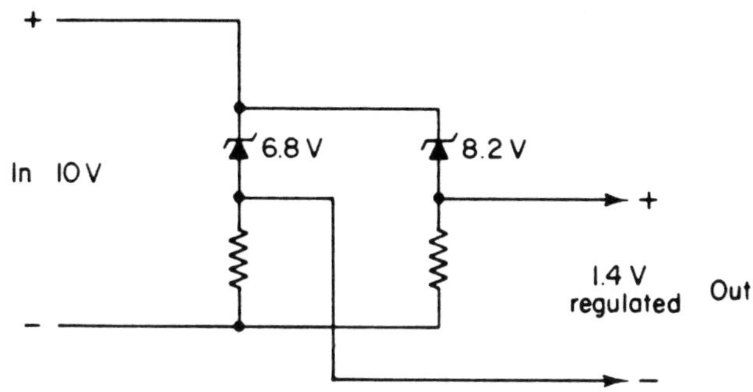

Figure 1-11. Using zeners for lower output voltages

arrangement provides good temperature compensation since both zeners tend to drift in the same direction with temperature changes, thus maintaining the difference voltage.

1.9.7 Using Zeners for an Adjustable Output

Zeners can be arranged to provide an adjustable regulated voltage as shown in Fig. 1–12. Any combination of zeners can be used to obtain the desired range of output voltage.

1.10 Adjustable-Shunt Regulators

Figure 1–13 shows the schematic and characteristics of an adjustable-shunt regulator (the classic Texas Instruments TI430). Such devices are also known as *programmable zeners* and are similar to three-terminal regulators (see Chapter 4) and voltage references (see Chapter 7). These regulators are a type of zener (in IC form). The IC in Fig. 1–13 is a three-terminal programmable shunt. (Shunt regulators are discussed in Chapter 3.)

The programmable zener of Fig. 1–13 has a thermal stability (or temperature coefficient, tempco) of 100 ppm/°C and can be used to replace other zeners of similar ratings. Programmable zeners have a sharp breakdown or avalanche point, as shown by the graphs. This particular zener is programmable over a voltage range of about 3 to 30 V and is capable of shunting up to 100 mA.

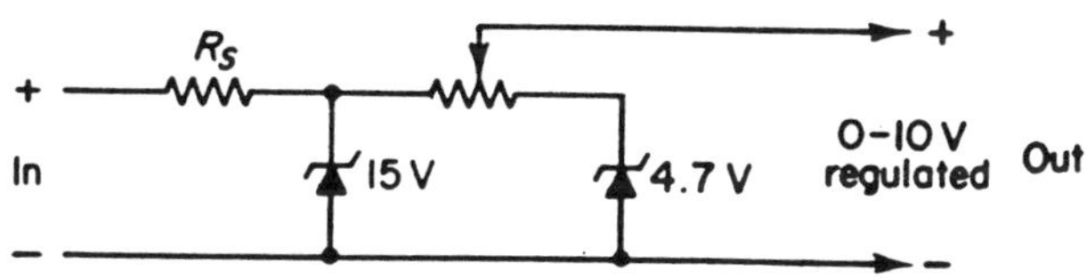

Figure 1-12. Using zeners for an adjustable output

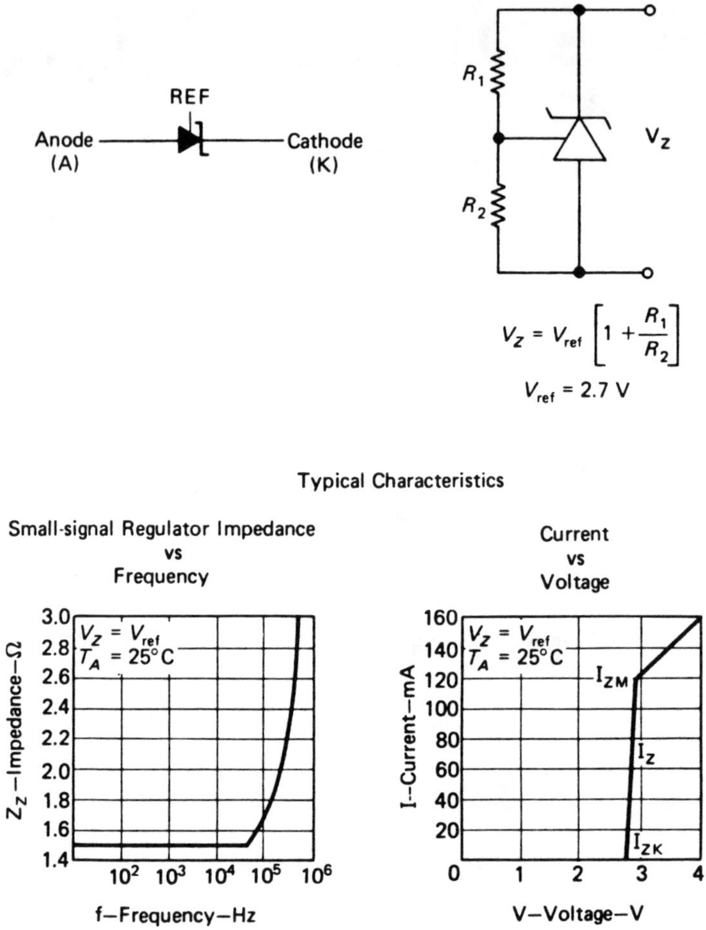

Figure 1–13. Schematic and characteristics of an adjustable shunt regulator

The zener is programmed by an external resistor network connected across the anode and cathode, with the reference terminal connected to the junction of R_1/R_2. The resistor values are selected so that the voltage developed across R_2 is 2.7 V for the desired programmed zener voltage.

The programmable zener is not limited to shunt-regulator functions. Figures 1–14 and 1–15 show the programmable zener in a variety of regulator functions. The figures also show the calculations necessary to find the values of external components in each application.

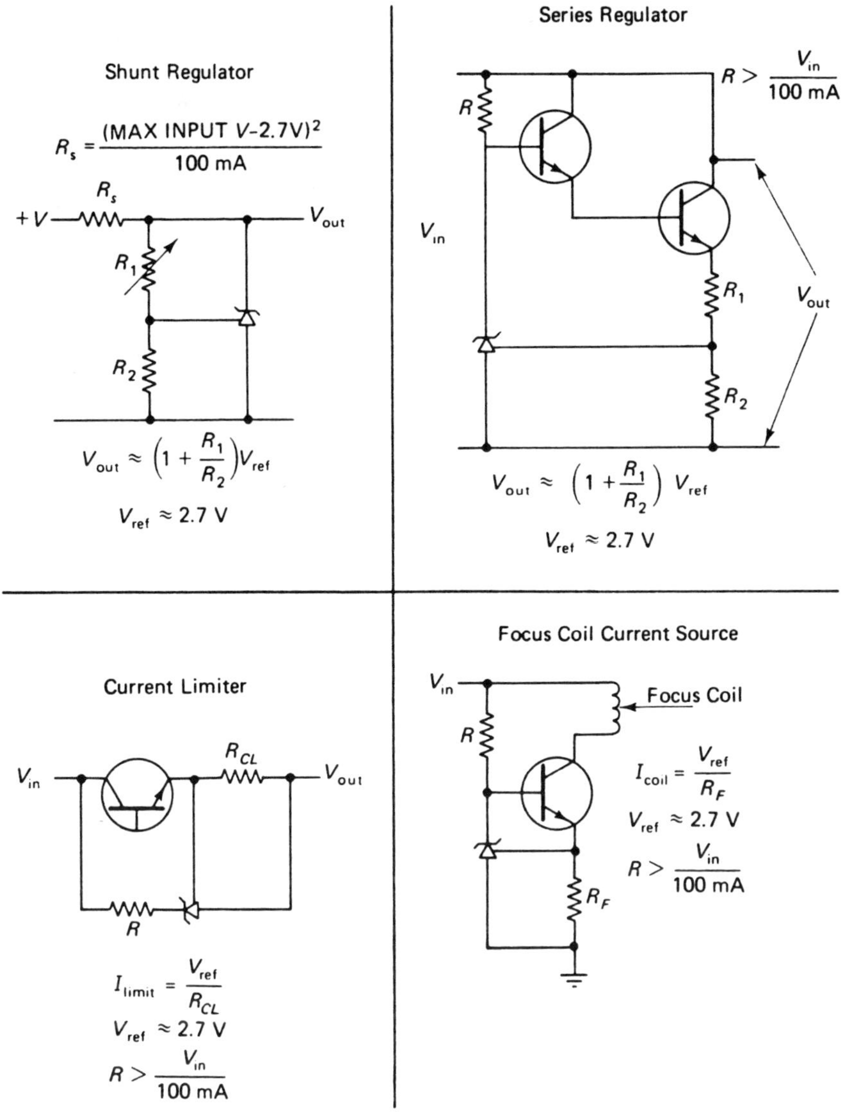

Figure 1–14. Applications for programmable zener

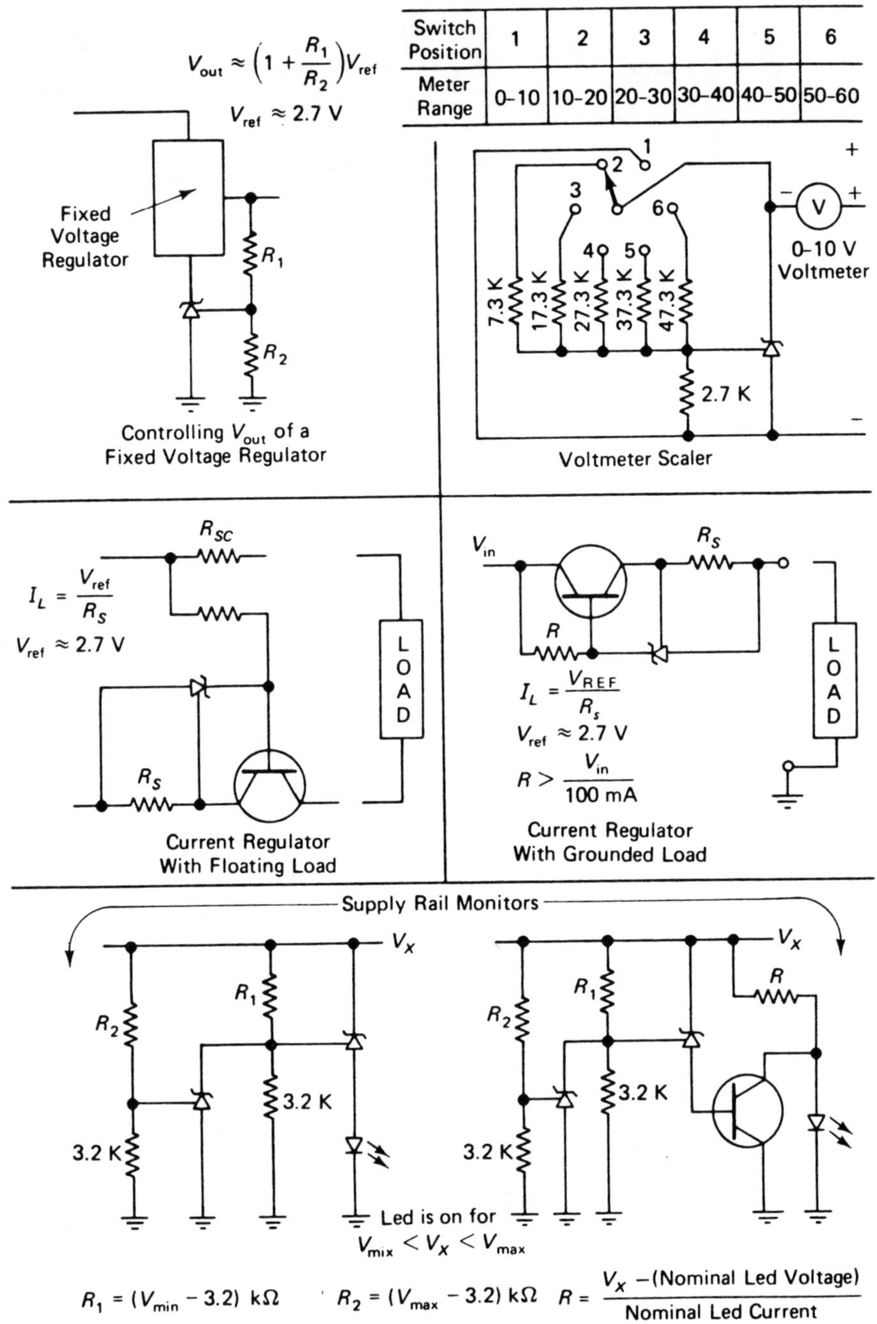

Switch Position	1	2	3	4	5	6
Meter Range	0–10	10–20	20–30	30–40	40–50	50–60

Figure 1-15. Additional applications for programmable zener

CHAPTER 2

Heat Sinks for Linear Power Supplies

This chapter is devoted to temperature-related design problems. One major disadvantage of a linear supply is the power dissipated by the series or shunt element. As discussed throughout the book, linear supplies contain at least one element (the series-pass transistor) that must pass the full load current. Such transistors can be part of a linear-regulator IC or external to the IC (where an IC operates as a control element). Either way, if the power dissipated by the series-pass element exceeds about 1 W, the elements must be provided with a heat sink. This also applies to zeners. Before we get into the details, let us review some basic temperature-related design problems for power supplies (and most other electronic equipment).

2.1 Temperature-Related Design Problems

There are two basic temperature-related problems in simplified linear-power-supply design. First, although data sheets specify component parameters (for transistors, diodes, rectifiers, ICs, etc.) at a given temperature, many of these parameters change with temperature. Because components rarely operate at the exact temperature shown on the data sheet, it is important to know the parameters at the actual operating temperature.

For example, in the case of transistors, three critical parameters (from a linear-supply design standpoint) are current gain, collector leakage, and power dissipation. To compound this problem, changes in parameters can affect transistor temperature (an increase in current gain or power dissipation results in a temperature increase).

Second, in addition to knowing the effects of temperature on parameters, it is important to know how heat sinks or component mounting can be used to offset the effects of temperature. For example, if a transistor (or diode, rectifier, IC) is used with a heat sink, or is mounted on a metal chassis that acts as a heat sink, an increase in temperature (from any cause) can be dissipated into the surrounding air.

The following paragraphs describe methods for approximating important parameters at temperatures other than those shown on data sheets. Methods for determining the proper dissipation characteristics of heat sinks are also discussed. (Temperature-related design problems for ICs are discussed in Section 2.3.)

2.1.1 Effects of Temperature on Transistor Collector Leakage

Collector leakage (I_{cbo}) increases with temperature. As a guideline, collector leakage doubles with every 10°C increase in temperature for germanium transistors and doubles with every 15°C increase for silicon transistors.

Collector leakage also increases with voltage applied at the collector. For example, a typical data-sheet leakage figure could be 2 μA at 25°C and 50 μA at 150°C. However, the 25°C figure is with a collector-base voltage of 30 V, while the 150°C figure is with 5 V. If the temperature is raised from 25°C to 150°C with 30 V at the collector, the collector leakage is about 500 μA.

As a result, always consider the possible effects of a different collector voltage when approximating collector leakage at temperatures other than those on the data sheet.

2.1.2 Effects of Temperature on Transistor Current Gain

Current gain (hfe) increases with temperature. As a guideline, current gain doubles when the temperature is raised from 25°C to 100°C for germanium transistors and doubles when the temperature is raised from 25°C to 175°C for silicon transistors.

It is obvious that silicon transistors are less temperature sensitive than germanium transistors. Data sheets usually specify a maximum operating temperature, or ambient temperature. If this temperature is not given or is unknown the guideline is this: Do not exceed 100°C for germanium transistors or 200°C for silicon transistors.

2.1.3 Effects of Temperature on Power Dissipation

The power-dissipation capabilities of a transistor (and diode, rectifier, or IC) must be carefully considered when designing any power-supply circuit. Of course, in small-signal circuits (not the series-pass transistor!), the power dissipation is usually less than 1 W, and heat sinks are not needed. In such circuits, the only concern is that the rated power dissipation (as shown on the data sheet) not be exceeded.

Here is a guideline: Do not exceed 90% of the maximum data-sheet power dissipation for small-signal circuits. This includes most of the circuits in linear supplies, except the series-pass element (transistor or IC), the rectifier diodes, and possibly the zeners.

As with other characteristics, manufacturers specify maximum power dissipation in a variety of ways on data sheets. Some manufacturers provide *safe-operating-*

area curves or graphs for temperature and/or power dissipation. Other manufacturers specify *maximum power dissipation,* in relation to a given ambient temperature or a given case temperature. Still others specify a *maximum junction temperature* or *maximum case temperature.*

2.1.4 Thermal Resistance

Transistors, rectifiers, and ICs designed for power-supply applications (or any power application) usually have some form of thermal resistance specified to show the power-dissipation capabilities. Thermal resistance can be defined as the increase in temperature of the component junction (with respect to some reference) divided by the power dissipated, or °C/W.

Power-transistor data sheets usually specify thermal resistance at a given temperature. This is also the case for many diodes and rectifiers. For each increase in temperature from the specified value, there is a change in the temperature-dependent characteristics of the component.

Because there is a change in temperature with changes in power dissipation of the component, the junction-to-ambient temperature also changes, resulting in a characteristic change. As a result, the component characteristics can change with ambient-temperature changes, and with changes produced by variation in power dissipation.

In power transistors, rectifiers, and ICs, thermal resistance is normally measured from the component junction to the case. This results in the term θ_{JC}, where the lowercase Greek letter theta (θ) indicates thermal resistance (because engineers like Greek letters!).

On those components where the case is bolted directly to the mounting surface with a built-in threaded bolt or stud, the term θ_{MB} (thermal resistance to mounting base) or θ_{MF} (thermal resistance to mounting flange) is used. These terms take into account only the thermal paths from junction to case (or mount). For power components in which the junction is mounted directly on a header or pedestal, the total internal thermal resistance from junction to case (or mount) varies from about 50°C/W to less than 1°C/W.

2.1.5 Thermal Runaway

The main problem in operating a transistor or IC near the maximum power limits is a condition known as thermal runaway. (Although thermal runaway can apply to diodes and rectifiers, they are generally not as affected by runaway, since they do not have current gain.)

When current passes through a transistor junction, heat is generated. If all the heat is not dissipated by the case (an impossibility), the junction temperature increases. In turn, this causes more current to flow through the junction, even though the voltage, circuit values, and so on remain the same. The increased current flow causes the junction temperature to increase even further, with a corresponding increase in current flow. If the heat is not dissipated by some means, the transistor burns out and is destroyed.

2.1.6 Operating without Heat Sinks

If a transistor or IC is not mounted on a heat sink, the thermal resistance from case to ambient air, θ_{CA}, is so large in relation to that from junction to case (or mount) that the total thermal resistance from junction to ambient air, θ_{JA}, is primarily the result of the θ_{CA} term.

Table 2–1 is a list of case-to-ambient thermal resistances for a few common transistor/IC cases (both old and new). As shown, heavy-duty cases such as TO–3 have a small temperature increase (for a given wattage) in comparison to such cases as the TO–5 (because heavy-duty cases dissipate the heat into the ambient air). Notice that the values shown in Table 2–1 are for 25°C and must be derated for ambient temperatures above 25°C.

The information in Table 2–1 can be used to approximate the maximum power dissipation of transistors (without heat sinks) when such information is not shown on the data sheet. For example, assume that a silicon transistor with a TO–5 case is used, and that the absolute maximum power dissipation (without a heat sink) must be found.

The case-to-ambient thermal resistance for a TO–5 case is 150. As discussed in Section 2.1.2, silicon transistors should not be operated above 200°C (under any circumstances). Assuming a 25°C ambient temperature, the transistor temperature should not be allowed to increase more than 175°C maximum (200 – 25 = 175). With a factor of 150 for the TO–5 case and a 175°C increase, the case must dissipate 1.17 W (175/150 = 1.17).

However, as discussed in Section 2.1.2, silicon-transistor current gain doubles when the temperature is raised from 25°C to 200°C. Assuming that the voltage remains constant, the maximum dissipation allowable is then cut in half to 0.585 W (or 585 mW). This is an absolute maximum figure, assuming a silicon transistor, TO–5 case, and an ambient temperature of 25°C. (For simplified design purposes, the 585-mW figure is safe if the case is mounted on a metal chassis, or other metal area, that can act as a heat sink.)

Table 2–1. Thermal resistance for common transistor/IC cases

Case	*θCA (°C/W)*
TO-3	30
TO-5	150
TO-8	75
TO-18	300
TO-36	25
TO-39	150
TO-46	300
TO-60	70
TO-66	60

2.1.7 Operating with Heat Sinks

After about 1 W, it becomes impractical to increase the size of the case to make the case-to-ambient thermal-resistance factor comparable to the junction-to-case factor. However, there are IC regulators with special TO–66 cases that can dissipate up to about 3 W. Except for these special circumstances, most power transistors, rectifiers, and ICs are designed for use with an external heat sink. Sometimes the mounting area serves as the heat sink. In other cases, a heat sink is attached to the case. Either way, the primary purpose of the heat sink is to increase the effective heat-dissipation area of the case, and to provide a low heat-resistance path from case to ambient.

Section 2.2 discusses the practical aspects of heat-sink design and selection. The following paragraphs describe the basic calculations involved.

To properly design (or select) a heat sink for a given application, the thermal resistance of both the component and heat sink must be known. For this reason, some power-transistor/IC data sheets specify the θ_{JA} that must be combined with the heat-sink thermal resistance to find the total power-dissipation capability.

Notice that some power-transistor/IC data sheets specify a maximum case temperature rather than θ_{JA}. As discussed in Section 2.1.11, maximum case temperature can be combined with heat-sink thermal resistance to find maximum power dissipation.

2.1.8 Heat-Sink Ratings

Commercial heat sinks are available for various case sizes and shapes (see Section 2.2). Such heat sinks are especially useful when the components are mounted such that there is little or no thermal conduction (heat transfer) to the PC board.

Commercial heat sinks are rated by the manufacturers in terms of thermal resistance, usually in terms of °C/W. When heat sinks involve the use of washers, the °C/W factor usually includes the thermal resistance between the case and sink, θ_{CA}. With a washer, only the sink-to-ambient, θ_{SA}, thermal-resistance factor is given. Either way, the thermal-resistance factor represents temperature increase (in °C) divided by wattage dissipated.

For example, if the heat-sink temperature rises from 25°C to 100°C when 25 W is dissipated, the thermal resistance is 75/25 = 3. This is listed on the data sheet as θ_{SA}, or simply 3°C/W.

All other factors being equal, the heat sink with the lowest thermal resistance (°C/W) is best. For example, a heat sink with 1°C/W is better than one with 3°C/W. Of course, the heat sink must fit the case and space around the case. Except for these factors, selection of a suitable heat sink should be no particular problem.

Table 2–2 is a brief selection guide to heat-sink manufacturers. No attempt has been made to provide a complete list. Likewise, the list covers only the basic case types of components commonly used in linear supplies (TO–66, TO–99, DIP).

Table 2–2. Commercial heat sink selection guide (*Raytheon Linear Integrated Circuits,* 1989, p. 9–74/9–75)

$\theta_{S\text{-}A}$*(°C/W)	Manufacturer/Series or Part Number
	TO-66 Package
0.31-1.0	Thermalloy — 6441, 6443, 6450, 6470, 6560, 6590, 6660, 6690
1.0 - 3.0	Wakefield — 641 Thermalloy — 6123, 6135, 6169, 6306, 6401, 6403, 6421, 6423, 6427, 6442, 6463, 6500
3.0 - 5.0	Wakefield — 621, 623 Thermalloy — 6606, 6129, 6141, 6303 IERC — HP Staver — V3-3-2
5.0 - 7.0	Wakefield — 690 Thermalloy — 6002, 6003, 6004, 6005, 6052, 6053, 6054, 6176, 6301 IERC — LB Staver— V3-5-2
7.0 - 10.0	Wakefield — 672 Thermalloy — 6001, 6016, 6051, 6105, 6601 IERC — LA, uP Staver — V1-3, V1-5, V3-3, V3-5, V3-7
10.0-25.0	Thermalloy — 6-13, 6014, 6015, 6103, 6104, 6105, 6117
	TO-99 Package
12.0 - 20.0	Wakefield — 260 Thermalloy — 1101, 1103 Staver — V3A-5
20.0 - 30.0	Wakefield — 209 Thermalloy — 1116, 1121, 1123, 1130, 1131, 1132, 2227, 3005 IERC — LP Staver — F5-5
30.0 - 50.0	Wakefield — 207 Thermalloy — 2212, 2215, 225, 2228, 2259, 2263, 2264 Staver — F5-5, F6-5
	Dual-Inline Package
20 30 32 34 45 60	Thermalloy — 6007 Thermalloy — 6010 Thermalloy — 6011 Thermalloy — 6012 IERC — LIC Wakefield — 650, 651

* All values are typical as given by manufacturer or as determined from characteristic curves supplied by manufacturer.

Staver Co., Inc.: 41-51 N Saxon Ave., Bay Shore, NY 11706
IERC: 135 W Magnolia Blvd., Burbank, CA 91502
Thermalloy: P.O. Box 34829, 2021 W Valley View Ln., Dallas, TX
Wakefield Engin Ind: Wakefield, MA 01880

2.1.9 Calculating Heat-Sink Capabilities

The thermal resistance of a heat sink can be calculated if the following factors are known: material, mounting provisions, exact dimensions, shape, thickness, surface finish, and color. Even if all these factors are known, the thermal-resistance calculations are approximate.

As a *very approximate* guideline, heat-sink thermal resistance in °C/W equals $\sqrt{1500/\text{area}}$, where the area (total area exposed to the air) is in square inches, material is $\frac{1}{8}$-inch thick, and the shape is a flat disk.

From a simplified design standpoint, it is better to accept the manufacturer's specifications for a heat sink. The heat-sink thermal resistance actually consists of two series elements: the thermal resistance from case to the heat sink (θ_{CS}) that results from conduction, and the thermal resistance from the heat sink to the ambient air (θ_{SA}) caused by convection and radiation.

2.1.10 Practical Heat-Sink Considerations

To operate a component at full power capability, there should be no temperature difference between the case and ambient air. This occurs only when the thermal resistance of the heat sink is zero, and the only thermal resistance is that between the junction and case. Although it is not practical to manufacture a heat sink with zero resistance, the greater the ratio θ_{JC}/θ_{CA}, the nearer the maximum power limit (set by θ_{JC}) can be approached.

When transistors are used with heat sinks, some form of electrical insulation must be provided between the case and heat sink (unless a grounded-collector circuit is used). Because good electrical insulators usually are good thermal insulators, it is difficult to provide electrical insulation without introducing some thermal resistance between the case and heat sink.

The best materials for electrical insulation of heat sinks are mica, beryllium oxide (Beryllia), and anodized aluminum. The properties of these three materials for case-to-heat-sink insulation of a TO–3 case are compared in Table 2–3.

For small, general-purpose transistors with a TO–5 case, a beryllium-oxide washer can be used to provide insulation between the case and a metal mounting surface. The use of a zinc-oxide-filled silicon compound (see Section 2.2.4) between the washer and mounting surface, together with a moderate amount of pressure from the transistor top, helps to decrease thermal resistance. If the transistor is mounted

Table 2–3. Comparison of insulating washers for head sinks

Material	*Thickness (in.)*	*(°C/W)*	*Capacitance (pf)*
Beryllia	0.063	0.25	15
Anodized aluminum	0.016	0.35	110
Mica	0.002	0.4	90

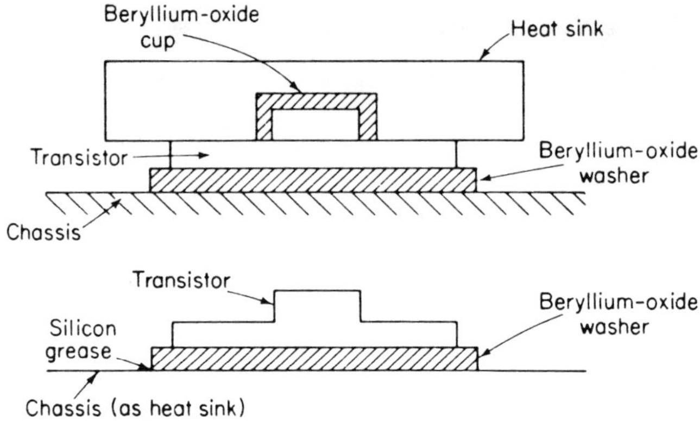

Figure 2–1. Typical mounting arrangements for transistor heat sinks

within a heat sink, a beryllium cup should also be used between the transistor and heat sink. Figure 2–1 shows both types of mounting. Section 2.2 describes mounting techniques for power transistors.

2.1.11 Calculating Power Dissipation

For simplified design, the collector voltage and current can be used to calculate power dissipation when a transistor is operated under steady-state conditions (as is the case in most linear-supply circuits). Actually, there are other currents that result in transistor power dissipation (collector-base leakage current, emitter-base current). However, these can be ignored, and the power dissipation (in watts) can be considered as the collector voltage times the collector current.

For a series-pass transistor, the difference between input and output voltages, multiplied by the current, determines the power dissipation. The *maximum input voltage,* the *minimum output voltage,* and the *maximum load current* are the critical design factors. For example, if the input voltage to a series-pass transistor can vary between 10 and 15 V, while the output is normally 8 V but can drop to 6 V when a maximum load of 3 A is drawn through the transistor, there is a possible differential of 9 V (15 – 6). This results in a dissipation of 27 W (9 × 3 = 27).

Once the power dissipation has been calculated, the maximum power dissipation capability must be found. Under the steady-state conditions of a linear power supply, the maximum dissipation capability depends on three factors: the sum of the series thermal resistances from the junction to ambient air, the maximum junction temperature, and the ambient temperature.

Following are some examples of how power dissipation can be calculated. Assume that we wish to find the maximum power dissipation of a transistor (in watts). The following conditions are specified: a maximum junction temperature of 200°C (typical for a silicon power transistor), a junction-to-case thermal resistance of

2°C/W, a heat sink with a thermal resistance of 3°C/W, and an ambient temperature of 25°C.

First, find the total junction-to-ambient thermal resistance: 2°C/W + 3°C/W = 5°C/W. Next, find the maximum permitted power dissipation (200°C – 25°C)/5°C/W = 35 W (maximum).

If the same transistor is used without a heat sink but under the same conditions and with a TO–3 case, the maximum power can be calculated as follows. First, find the total junction-to-ambient thermal resistance: 2°C/W + 30°C/W = 32°C/W. Next, find the maximum permitted power dissipation (200°C – 25°C)/32°C/W = 5 W (approximately).

Some power-transistor data sheets specify a *maximum case temperature* rather than a maximum junction temperature. Assume that a maximum case temperature of 130°C is specified instead of a maximum junction temperature of 200°C. In that event, subtract the ambient temperature from the maximum permitted case temperature: 130°C – 25°C = 105°C. Then divide the case temperature by the heat-sink thermal resistance: 105°C/3°C/W = 35 W maximum power.

2.2 Heat-Sink Mounting

Proper mounting procedures must be followed to minimize the *interface* thermal resistance between component (transistor, rectifier, IC, etc.) package or case and the heat sink. Proper mounting provides cooler and more reliable operation of the components.

This section describes mounting techniques of metal-packaged power transistors (and other power devices such as rectifiers and IC regulators). Included in this section are discussions concerning preparation of the mounting surface, use of thermal compounds, and fastening techniques. Typical interface thermal resistance is given for a few packages commonly found in linear-supply circuits.

As discussed, the total thermal resistance junction-to-ambient is the sum of the individual thermal resistances (junction-to-case, case-to-heat-sink, heat-sink-to-ambient). The junction-to-case and sink-to-ambient resistances are characteristics of the component and heat sink, respectively. So only the case-to-sink resistance can be changed, depending on mounting techniques.

The case-to-sink resistance is often referred to as the interface thermal resistance, sometimes listed as $R\theta_{CS}$. No matter what it is called, the lowest value of interface thermal resistance is best, since this produces the lowest junction-to-ambient resistance. In the following sections we discuss practical methods to get the lowest junction temperature under all mounting conditions.

2.2.1 Interface Thermal-Resistance Values

Table 2–4 shows the interface thermal resistance (in °C/W) for a number of diode and transistor/IC cases. Table 2–4 also shows recommended hole and drill sizes as well as torque for the mounting nuts or screws.

Table 2–4. Interface thermal resistance for a number of diode and transistor cases

Package type and data			Interface thermal resistance (°C/W)					
JEDEO outline number	Description	Recommended hole and drill size	Torque in-lb	Metal-to-metal		With insulator		
				Dry	Lubed	Dry	Lubed	Type
DO-4	10-32 stud 7/16 hex	0.118, No. 12	15	0.41	0.22	1.24	1.06	3-mil mica
DO-5	1/4-28 stud 11/16 hex	0.25, No. 1	30	0.38	0.20	0.89	0.70	5-mil mica
DO-21	Pressfit, 1/2		*	0.15	0.10	–	–	–
TO-3	Diamond	0.14, No. 28	*	0.20	0.10	1.45	0.8	3-mil mica
						0.8	0.4	2-mil mica
						0.4	0.35	Anodized aluminum
TO-66	Diamond	0.14, No. 28	*	–	0.50			
TO-83	1/2-20 stud	0.50, 0.50	130	–	0.10			

*Can be tapped for 10-24 machine screw

Notice that the interface thermal resistance changes quite drastically for different mounting conditions. For example, assume that a TO–3 case is involved. If the case is mounted (on the heat sink or metal surface) with an insulator but no thermal compound or lubrication, the interface thermal resistance is 1.45°C/W. If a thermal compound is used, the resistance drops to 0.8°C/W. If circuit conditions make it possible to eliminate the insulator, the thermal resistance drops to 0.1°C/W (with thermal compound) or 0.2°C/W (without compound).

2.2.2 Fastening Techniques

The various types of packages shown in Table 2–4 require different fastening techniques. Mounting details for stud, flat-base, press-fit, and disk-type components are shown in Figs. 2–2, 2–3, 2–4 and 2–5, respectively. The following notes supplement the illustrations.

With any of the mounting schemes, the screw threads should be free of grease, to prevent inconsistent torque readings when tightening nuts. Maximum allowable torque should always be used to reduce thermal resistance. However, care must be exercised not to exceed the torque rating of parts. Excessive torque applied to disk- or stud-mounted parts (Figs. 2–2 and 2–5) could cause damage to the semiconductor die.

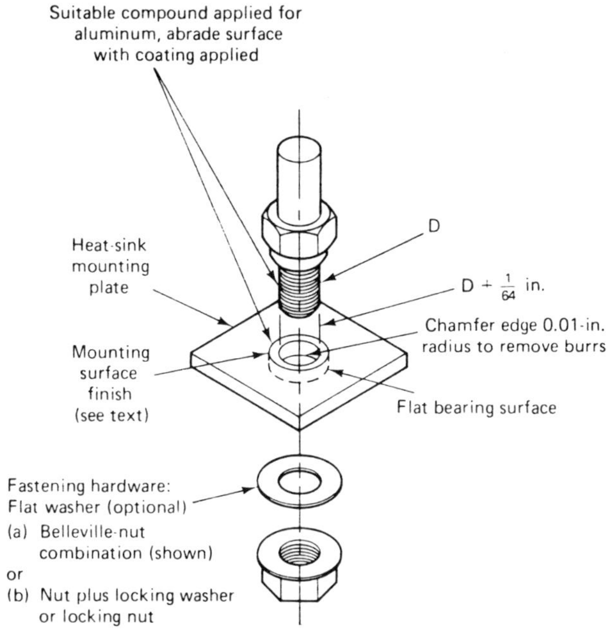

Figure 2–2. Mounting details for stud-mounted components

To prevent galvanic action from occurring when components are used with aluminum heat sinks in a corrosive atmosphere, many devices are nickel or gold plated. Take precautions not to mar the surface.

With press-fit components (Fig. 2–4), the hole edge must be chamfered as shown to prevent shearing off the knurled edge of the component during press-in. The pressing force should be applied evenly on the shoulder ring to avoid tilting or canting of the device case in the hole during the pressing operation. Also, thermal compound should be used to ease the component into the hole.

Typically, the pressing force varies from 250 to 1000 lb, depending on the heat-sink material. Recommended hardnesses for typical heat-sink materials are as follows: copper, less than 50 on the Rockwell F scale; aluminum, less than 65 on the Brinell scale. A heat sink as thin as ⅛ inch may be used, but the thermal resistance increases in proportion to the reduction in contact area. A thin chassis requires the addition of a backup plate.

With the disk-type mounting (Fig. 2–5), a self-leveling type of mounting clamp is recommended to assure parallel contacts and even distribution of pressure on each contact area. A swivel-type clamp or a narrow leaf-spring in contact with the heat sink is usually acceptable.

The clamping force should be applied smoothly, evenly, and perpendicular to the disk-type package to prevent deformation of the device, or the sink-mounting surfaces, during installation. The spring used should provide a mounting force

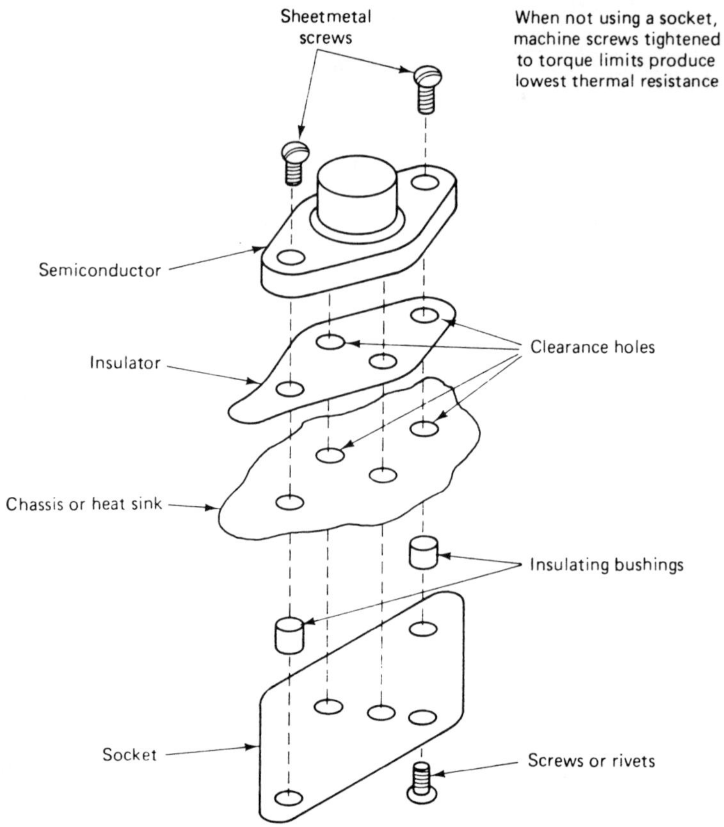

Figure 2–3. Mounting details for flat-base-mounted components

within the range recommended by the component manufacturer. Typical clamping forces for disk-type components are 800 to 2000 lb.

Installation of a disk-type device between two heat sinks should be done so as to permit one heat sink to move with respect to the other. Such movement prevents stresses from developing because of thermal expansion, which could damage the component.

When two or more components are to be operated electrically in parallel, one of the heat sinks can be common to both (or all) components. Individual heat sinks must be provided against the other mounting surfaces of the components so that the mounting force applied in each case is independently adjustable.

2.2.3 Preparation of Mounting Surface

In general, the heat sink should have a flatness and finish comparable to that of the component. For the typical experimenter or hobbyist, the heat-sink surface is satisfactory if it appears flat against a straightedge and is free of any deep scratches.

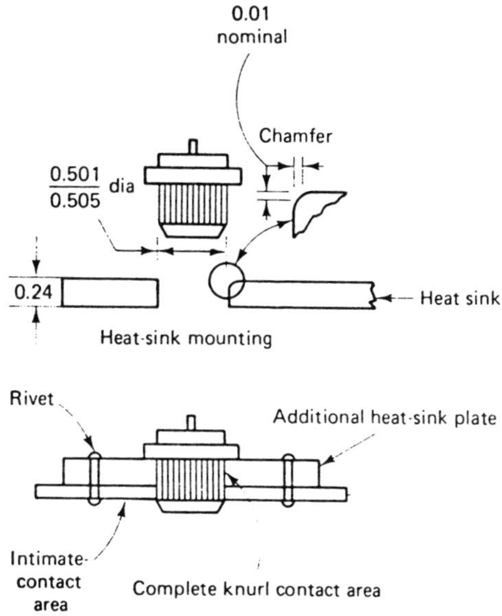

Figure 2–4. Mounting details for press-fit components

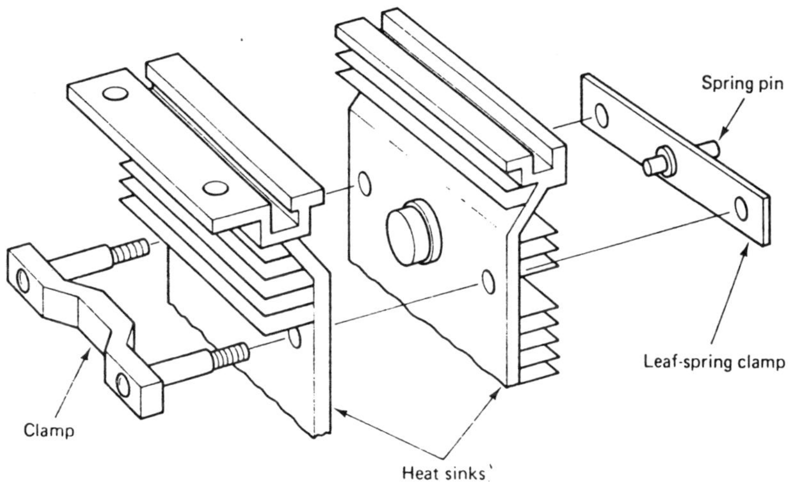

Figure 2–5. Mounting details for disk-type components

Of course, in commercial power supplies, it is necessary to measure the actual flatness with special tools and indicators.

Most commercially available case or extruded heat sinks require spot-facing. In general, milled or machine surfaces are satisfactory if prepared with tools in good working condition.

The surface must be free from all foreign material, film, and oxide (freshly bared aluminum forms an oxide layer in a few seconds). Unless the heat sink is used immediately after machining, the mounting area should be polished with No. 000 steel wool, then rinsed with acetone or alcohol, and coated immediately with thermal grease or compound.

Many aluminum heat sinks are black-anodized for appearance, durability, performance, and economy. Anodizing is an electrical and thermal insulator that offers resistance to heat flow. As a result, anodizing should be removed from the mounting area.

Another aluminum finish is irridite (chromate acid dip), which offers low resistance because of the thin surface. For best performance, the irridite finish must be cleaned of oils and films that collect when heat sinks are manufactured and stored.

Some heat sinks are painted after manufacture. Since paint of any kind has a high thermal resistance (compared to metal), paint must be removed from the heat-sink surface where the component is attached.

2.2.4 Thermal Compounds

Thermal compounds (also called *joint compounds* or *silicon greases*) are used to fill air voids between the mating surfaces, so as to improve contact between the component and heat sink. A typical compound has a resistivity of about 60°C-in./W, compared to about 1200°C-in./W for air. As a result, the thermal resistance of voids, scratches, and imperfections filled with a compound or grease is about one-twentieth of the original resistance.

Thermal compounds are a formulation of fine zinc particles in a silicon oil that maintains a greaselike consistency with time and temperature. There are two common methods for applying the compounds. With one technique, the compound is applied in a very thin layer with a spatula or lintless brush, and then the surface is wiped lightly to remove excess material. The other technique involves applying a small amount of pressure to spread the compound. After the mounting is complete, any excess compound is then wiped away using a cloth moistened with acetone or alcohol.

Some recommended thermal compounds (for linear-supply heat sinks) follow:

Astrodyne—Conductive Compound 829
Dow Corning—Silicon Heat Sink Compound 340
Emerson & Cuming, Inc.—Eccotherm, TC-4
General Electric—Insulgrease
George Risk Industries—Thermal Transfer Compound XL500
IERC—Thermacote
Wakefield—Thermal Compound Type 1201

2.3 Heat Sinks for ICs

The basic rules for linear-supply ICs regarding power dissipation and thermal considerations are essentially the same as those for discrete transistors and rectifiers, as discussed in Section 2.1.

The maximum allowable power dissipation (usually specified as PD or *maximum device dissipation* on IC data sheets) is a function of the maximum storage temperature T_S, the maximum ambient temperature T_A, and the thermal resistance from the semiconductor chip to case. The basic relationship is $P_D = (T_S - T_A)$/thermal resistance.

Every IC data sheet does not necessarily list all of these parameters. It is quite common to list only the maximum power dissipation for a given ambient temperature and then show a *derating factor* in terms of maximum power decrease for a given increase in temperature.

For example, a typical IC might show a maximum power dissipation of 110 mW at 25°C, with a derating factor of 1 mW/°C. If the IC is operated at 100°C, the maximum power dissipation is 100 – 25, or 75°C increase; 110 – 75 = 35 mW.

In the absence of specific data-sheet information, the following typical temperature characteristics can be applied to the basic IC package types. No IC should have a temperature in excess of 200°C.

Ceramic Flat Pack

Thermal resistance: 140°C per watt
Maximum storage temperature: 175°C
Maximum ambient temperature: 125°C

TO–5 Style (Metal Can) Package

Thermal resistance: 140°C per watt
Maximum storage temperature: 200°C
Maximum ambient temperature: 125°C

DIP (Ceramic) Package

Thermal resistance: 70°C per watt
Maximum storage temperature: 175°C
Maximum ambient temperature: 125°C

DIP (Plastic) Package:

Thermal resistance: 150°C per watt
Maximum storage temperature: 85°C
Maximum ambient temperature: 75°C

2.3.1 Working with Regulator ICs

Most present-day ICs require low power (typically less than 1 W) and can be operated without heat sinks. However, there are some linear-supply ICs (regulators) that must be used with a heat sink (either an external heat sink or by direct contact with a metal surface). Power ICs generally use some form of metal package.

The data sheets for these power ICs usually list sufficient information to select the proper heat sink. Also, the data sheets or other literature often recommend how to mount the power IC. Always follow the IC manufacturer's recommendations. In the absence of such data, and to familiarize the reader with the terms, the following sections summarize considerations for power ICs.

2.3.2 Maximum Power Dissipation

From a simplified design standpoint, an IC regulator (see Chapter 4) is a complete, predesigned, functioning circuit that cannot be altered in regard to power dissipation. That is, if the power-supply voltages, input signals, output loads, and ambient temperature are at the recommended levels, the power dissipation will be well within the capabilities of the IC.

With the possible exception of the data required to select or design heat sinks, the linear-supply designer need only follow the data sheet recommendations. There are many worked-out design examples for linear supplies, using off-the-shelf ICs, in Chapter 7. These include calculations for operating with heat sinks and/or for operating the IC at a lower power-dissipation capacity because of increased temperature (derating factors).

2.3.3 Thermal Resistance for Power ICs

ICs designed for power applications usually have some form of thermal resistance specified to indicate the power-dissipation capability, rather than a simple maximum device dissipation, or P_D. Thermal resistance can be defined as the increase in temperature of the semiconductor material, with regard to some reference, divided by the power dissipated, or °C/W.

Power-IC data sheets often specify thermal resistance at a given temperature. For each increase in temperature from this specified value, there is a change in the temperature-dependent characteristics of the IC. Because there is a change in temperature with changes in power dissipation, the semiconductor chip temperature also changes, resulting in a characteristic change.

IC characteristics can change with ambient-temperature changes and with changes produced by variations in power dissipation (such as changes in load current passing through the IC). As discussed in the following paragraphs, most power ICs have circuits to offset the effects of temperature.

In power ICs, thermal resistance is normally measured from the semiconductor chip to the case. On those ICs where the cases are bolted directly to the mounting surfaces with a threaded bolt or stud (similar to that shown in Fig. 2–2), the thermal resistance is measured from the chip to the mounting stud or flange.

2.3.4 Thermal Runaway in Power ICs

Power ICs (and some other ICs) are subject to the problems of thermal runaway (as discussed for transistors in Section 2.1.5). However, unlike discrete transistors and rectifiers, most ICs have built-in circuits to prevent thermal runaway. The usual arrangement is to place a diode in the reverse-bias circuit for one or more of the transistors in the IC. The diode is fabricated on the same semiconductor chip as the transistors and thus has the same temperature characteristics.

The thermal-protection circuit is arranged so that the reverse bias is increased (by a decrease in diode resistance) when there is an increase in temperature. When temperature increases because of an increase in internal transistor current (or vice versa), the diode resistance changes and increases the reverse bias. This offsets the initial change in current caused by temperature changes.

Many different temperature-compensation circuits have been developed by IC manufacturers. However, the IC user need not be concerned as long as the data-sheet limits are observed, because the circuit is already designed in the IC.

2.3.5 Thermal and Excess-Current Shutdown

Most linear-supply ICs (such as those described in Chapters 4 and 7) have a built-in thermal-shutdown circuit. If the IC temperature reaches a given value (typically 175°C), the circuit turns the IC off. Also, as described in Chapters 4 and 7, most regulator ICs have a provision for turnoff in case of excessive current flow through the IC. However, such turnoff is controlled by components external to the IC.

2.3.6 IC Heat-Sink Calculations

The calculations for IC heat sinks and power dissipation are essentially the same as for transistors and rectifiers described in Section 2.1; they are not repeated here. However, where there are a large number of ICs (as well as transistors and rectifiers) operated in a confined area, heat sinks may not do the job. The linear-supply designer may want to consider the use of *forced air* (such as from a fan or blower) to keep the ambient air within an acceptable tolerance. Forced air is often used in commercial power supplies where large operating currents are involved.

CHAPTER 3

Discrete Feedback-Regulator Basics

Basic feedback-regulator circuits include shunt voltage regulators, high-current regulators, constant-current (CC) regulators, series voltage regulators, overload protection for regulators, and feedback regulation (for both output and input control). Although the circuits described here are in discrete form, the same basic principles apply to the IC regulators, and combination discrete/IC regulators, described throughout the book.

3.1 Extending Zener Regulation

The voltage-control ability of a zener (see Chapter 1) can be increased if the zener is used to control the operating point of a transistor or group of transistors. There are two basic types of transistor regulators using zeners: shunt and series. The shunt regulator is placed across the power-supply output, while the series regulator is in series with the output.

Although some zeners can handle heavy currents and dissipate large amounts of power, it is usually less expensive to extend the current rating by means of a power transistor. In either the shunt or the series circuit, the power transistor is mounted on a heat sink (see Chapter 2) to dissipate maximum power.

With either shunt or series, the emitter-base junction of the power transistor is held at some fixed value by means of a low-power zener. The power transistor (typically silicon) emitter remains within about 0.5 V of the zener voltage, despite changes in input voltage or output (load) current. If a germanium transistor is used (not typical), the emitter remains within about 0.2 V of the zener voltage.

Although popular at one time, the shunt circuit has generally been replaced by the series circuit. The shunt regulator has two disadvantages: more power consumption (which is lost to the load) and a continuous load on the basic power supply. That is, the shunt circuit draws a continuous or *quiescent current,* even when no current is being delivered to the load. For these reasons, we concentrate on the series linear regulator here, although we do describe some of the basic shunt regulators, many of which are still in use.

3.2 Shunt Voltage Regulators

Figure 3–1 shows the simplest form of shunt regulator using a transistor. Transistor Q_1 appears across the power-supply output as a variable-bleeder resistor, with current flowing between emitter and collector. Base current flows through zener CR_1. Both of these currents, as well as the load current, flow through the series resistor R_1.

If the power-supply load should increase, more current is drawn through R_1 and the supply output voltage tends to drop. Under these conditions, less current is drawn through CR_1, the forward bias on Q_1 is lowered, and less emitter-collector current is drawn from the power supply. This causes a lower drop across R_1 and tends to increase the power-supply output voltage, thus offsetting the initial drop in voltage. The opposite occurs if the load decreases.

3.2.1 Cascade Shunt Regulator

Figure 3–2 shows the basic cascade-shunt-regulator circuit. Such circuits are used when the power supply is subject to large current changes. The cascade circuit increases the effectiveness of the basic shunt linear regulator.

Transistors Q_1 and Q_2 are placed across the power-supply output and act as variable resistors. Base current for Q_1 flows through CR_1, while base current for Q_2 flows through R_2. The voltage across R_2 is determined by current flowing through R_2 and Q_1. All of these currents, as well as load current, flow through the series resistor R_1.

As the load decreases, less current is drawn through R_1 and the output voltage rises. More current is drawn through CR_1 and through the base-emitter of Q_1. This raises the forward bias of Q_1, causing more emitter-collector current to be drawn from the power supply. Also, more current is passed through emitter resistor R_2, causing the voltage drop across R_2 to increase.

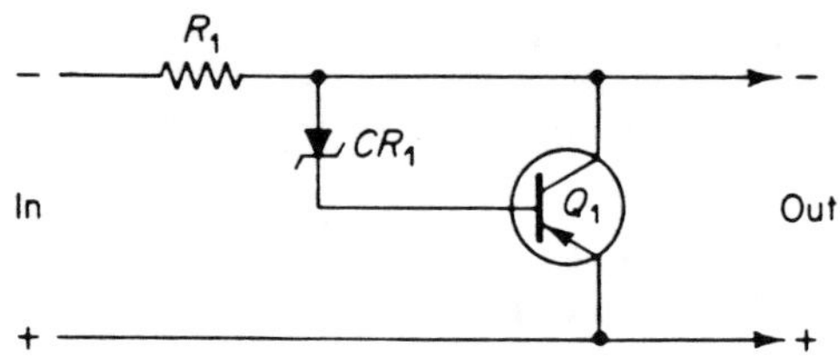

Figure 3–1. Basic shunt regulator

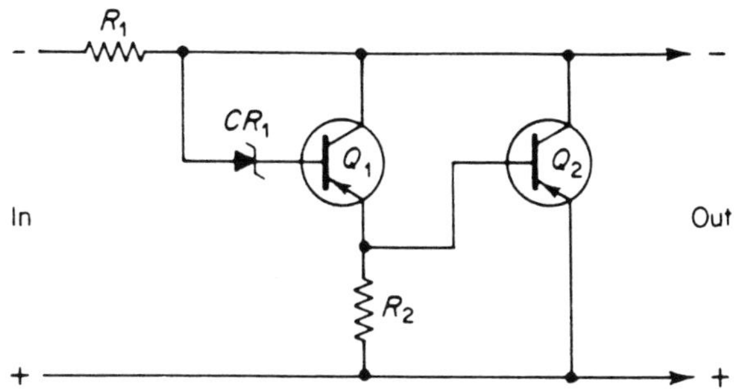

Figure 3–2. Basic cascade shunt regulator

When the drop across R_2 increases, the forward bias on Q_2 increases and more Q_2 emitter-collector current is drawn from the power supply. The increased current through Q_1 and Q_2 causes a higher drop across R_1, which tends to decrease the power-supply output voltage and thus offset the initial rise in voltage. The opposite occurs if the load increases.

3.2.2 Shunt Regulator for Voltages Higher than Zener Voltage

Figure 3–3 shows a shunt-regulated linear supply for outputs higher than the zener voltage. Neglecting the effect of R_S, or assuming that the source voltage is at the junction of R_S and R_1, the output voltage is determined by the R_1/R_2 ratio: output voltage = $(R_1 + R_2)/R_2$.

For example, if R_1 and R_2 are of the same value, the output voltage is twice that of the zener voltage.

Resistor R_3 compensates for variations in the supply to the regulator circuit. A high resistance of R_3 results in overcompensation, while an R_3 with too low a resistance produces undercompensation. (The value of R_3 is usually found by trial and error, using a variable resistance.)

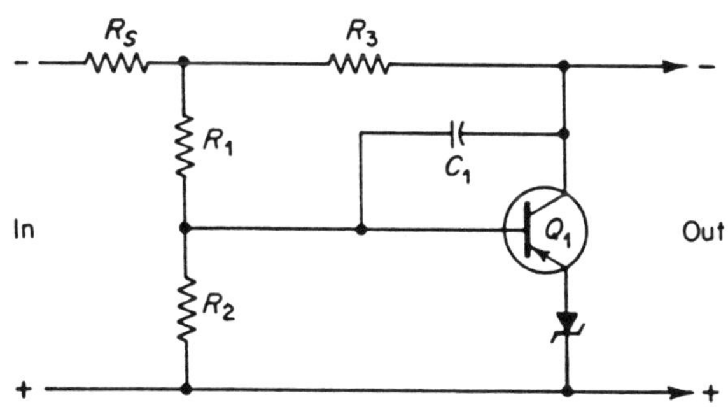

Figure 3–3. Shunt-regulated supply for voltages higher than the zener

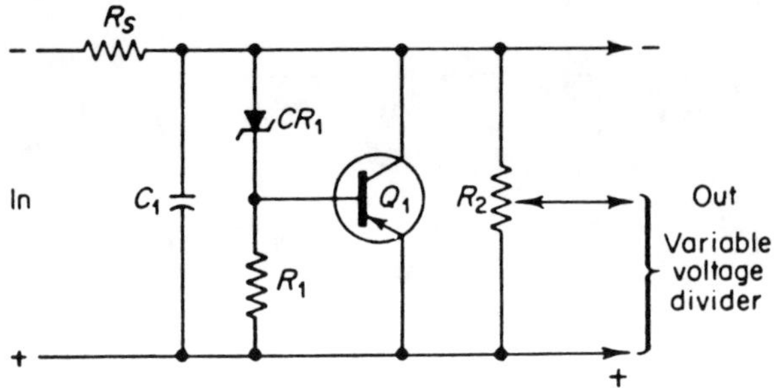

Figure 3–4. Shunt-regulated supply for voltages lower than the zener

3.2.3 Shunt Regulator for Voltages Lower than Zener Voltage

Figure 3–4 shows a shunt-regulated supply for outputs lower than the zener voltage. Potentiometer R_2 acts as a variable voltage divider and sets the regulated output voltage. The forward bias on Q_1 is determined by the voltage drop across R_1 (or the current through R_1).

If the supply voltage tends to increase, more current flows through CR_1 and R_1, causing an increase in forward bias on Q_1. The increase in Q_1 collector-emitter current passing through R_S causes a greater drop across R_S, offsetting the initial rise in supply voltage. The opposite occurs if the supply voltage decreases.

3.3 Series Voltage Regulators

Figure 3–5 shows the basic series voltage regulator. Transistor Q_1 is in series with the power-supply output (and possibly a fixed series resistance) and acts as a variable series resistor, with current flowing between the collector and emitter. (Q_1 is the *series-pass transistor* referred to throughout the book.)

Current also flows through resistor R_1 and zener CR_1 to establish a voltage at the base of Q_1. The base voltage remains fixed in relation to one terminal of the power supply (the negative terminal in this case) but varies in relation to the other terminal.

If the power-supply load increases, more current is drawn through the Q_1 emitter-collector resistance and the output voltage drops. Under these conditions, the Q_1 emitter voltage drops, but the base remains constant, increasing the forward bias on Q_1 (an NPN in this case). The increase in forward bias on Q_1 decreases the Q_1 emitter-collector resistance, raising the output voltage to offset the initial drop. The opposite occurs if the load decreases.

3.3.1 Series Voltage-Regulator Design Considerations

The following paragraphs summarize the design considerations for a discrete-transistor series linear regulator. Notice that the series-pass transistor will probably require a heat sink (see Chapter 2) if the power dissipated by the transistor exceeds 1 W.

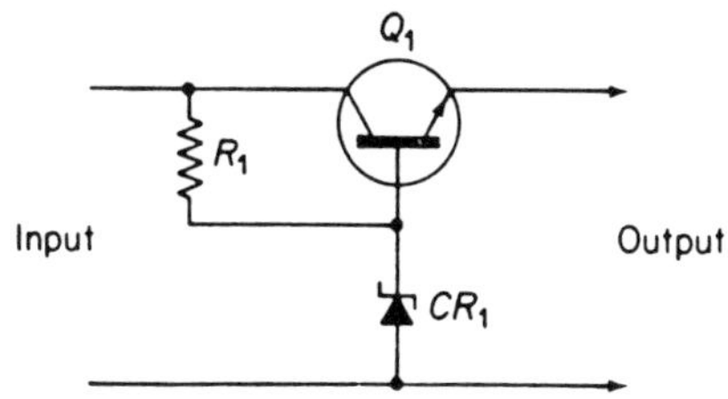

Output voltage: p–n–p silicon ≈ Zener − 0.5 V
p–n–p germanium ≈ Zener − 0.2 V
n–p–n silicon ≈ Zener + 0.5 V
n–p–n germanium ≈ Zener + 0.2 V

Maximum voltage, power, and current ≈ Q_1 ratings; see text
Minimum input voltage ≈ 1–2 V + Zener
Current thru R_1 ≈ 5% of total load current (at minimum input value)
Zener power rating ≈ 0.1 × total power output

$$R_1 \text{(in ohms)} \approx \frac{\text{(minimum input voltage − Zener voltage)}}{\text{(maximum output current} \times 0.5)}$$

$$\text{Maximum current thru } R_1 \approx \frac{\text{(maximum input voltage − Zener voltage)}}{R_1 \text{ resistance}}$$

R_1 (in watts) ≈ (maximum current thru R_1)^2 × R_1 resistance
Zener maximum current rating ≈ maximum current thru R_1

Figure 3–5. Basic series voltage regulator

3.3.2 Maximum Voltage, Current, and Power Dissipation

The input voltage, current, and power-dissipation characteristics of the series regulator are set by the series-pass transistor Q_1 (see Fig. 3–5), not by the zener. For example, maximum input voltage to the circuit is set by the maximum collector voltage of Q_1. The maximum current and power ratings of Q_1 fix the maximum load-current and power-dissipation capabilities of the circuit.

3.3.3 Output Voltage

The output voltage of the series regulator is set by the zener and is within 0.5 V (for silicon transistors) or 0.2 V (for germanium transistors) of the zener voltage. If a PNP is used, the output is 0.2 or 0.5 V below the zener value. If an NPN is used, the output is 0.2 or 0.5 V above the zener value.

3.3.4 Minimum Input Voltage

The minimum voltage capability of the regulator is also set by the zener voltage. As a rule, the minimum voltage (at the collector) must be at least 1 V, and preferably 2 V, greater than the zener voltage (at the base).

3.3.5 Base Resistor

The value of base resistor R_1 is selected to provide the optimum base current. As a first trial value, R_1 should be selected to provide about 5% of the total load current when the input voltage is at minimum.

3.3.6 Zener Diode

The zener should be selected on the basis of zener voltage and current (or power capability). Because the output voltage is close to the zener voltage, this sets the zener value. Although the output current and power dissipation of the series regulator depend (primarily) on the series-pass transistor capability, it is usually not practical to design any suitable series regulator with currents greater than 10 times the maximum rated zener current.

3.3.7 Design Example: Series Voltage Regulator

Assume that the circuit of Fig. 3–5 is to provide 32 W of power (8 V at 4 A maximum). The input voltage varies from 12 to 15 V. A silicon PNP transistor capable of handling 35 W is available (with heat sink).

Verify that the transistor is capable of handling the full voltage, current, and power. The maximum voltage applied is 15 V. Any transistor capable of 35-W dissipation will probably accept 15 V at the collector. However, this should be verified by reference to the data sheet. The same is true of maximum current, 4 A in this case.

Notice that the transistor need not dissipate the full power, since the maximum emitter-collector differential is 7 V (15 V input, less 8 V output). In theory the transistor should not dissipate more than 28 W (7 V × 4 A). So the 35-W capability provides a good safety margin.

Because the output voltage must be 8 V and the transistor is silicon PNP, the zener voltage is 8 + 0.5, or 8.5 V. (If the transistor is NPN as shown, the zener voltage is 8 – 0.5, or 7.5 V.)

The minimum input voltage is 8.5 + 1, or 9.5 V (or, preferably, 8.5 + 2, or 10.5 V) or greater. So the 11-V minimum input voltage presents no problem.

As a first trial value for base resistor R_1, use a resistance that produces a current at the base equal to 5% of the maximum output current when the input voltage is minimum, or $(11 - 8.5)/(4 \times 0.05) = 12.5$ ohms. Use the nearest standard of 15 or 18 ohms.

The wattage rating for R_1 is based on maximum current when the input voltage is maximum: $I = (15 - 8.5)/15 = 0.433$ A; $P = (0.433)^2 \times 15 = 2.81$ W. Use the nearest standard of 5 W.

The current through R_1 is divided between the base-emitter junction and the zener. Because this current is 0.433 A maximum, the zener must also be capable of 0.433 A, or 3.68 W (8.5×0.433). Use the nearest standard of 5 or 10 W for the zener.

Notice that the zener will probably never pass the full 0.433 A, because part of the current passes through the emitter-base junction. However, the zener rating should be based on maximum possible current (as a safety factor).

3.4 High-Voltage Linear Regulation

Figure 3–6 shows a discrete zener/transistor linear regulator circuit for high voltages. Although solid-state circuits are not usually associated with high voltages, it is possible to use zener/transistor circuits to regulate high-voltage linear supplies.

Transistor Q_1 is in series with the supply output and acts as a variable series resistor (as described in Section 3.3). Current also flows through R_1 to establish a voltage at the base of Q_1. This voltage is determined by the current flowing through R_1 and through the emitter-collector circuit of Q_2.

When the forward bias of Q_1 is decreased, the Q_1 emitter-collector resistance in series with the power supply is raised. This causes a higher voltage drop across the emitter-collector resistance and lowers the supply output voltage. The emitter-collector current of Q_2 is determined by the forward-bias voltage across R_2. This voltage is determined by the emitter-collector current of Q_3. The emitter of Q_2 is held at a fixed voltage by the action of CR_1.

Transistors Q_3 and Q_4 establish a control voltage for the regulator circuit. The base of Q_4 is held at a fixed voltage by CR_2. The emitter-collector current of Q_4 remains constant, and the emitter of Q_4 remains at a fixed voltage. Because the emitter of Q_3 is connected directly to the emitter of Q_4, the emitter of Q_3 also remains at a fixed voltage. However, the voltage at the base of Q_3 varies with the supply output voltage.

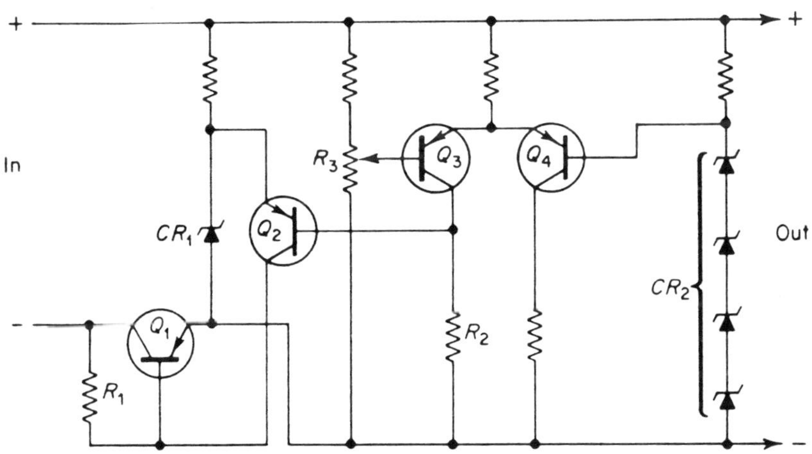

Figure 3–6. Discrete zener/transistor regulator for high voltages

If the supply output voltage changes, the forward bias on Q_3 changes, causing a change in the voltage drop across R_2. This changes the forward bias on Q_2, causing a change in emitter-collector current. Because the emitter-collector current of Q_2 passes through R_1, the voltage drop across R_1 changes, making the base of Q_1 more or less negative.

The change in Q_1 base voltage changes the Q_1 emitter-collector resistance in series with the supply output. In turn, this causes a higher or lower voltage drop across the emitter-collector resistance and changes the output voltage to offset the initial rise or fall in voltage.

The supply output voltage is set by adjustment of R_3, which determines the bias voltage on the base of Q_3. The supply is adjusted by connecting a voltmeter across the supply output terminals (usually with the load connected) and setting R_3 for the desired voltage.

3.5 High-Current Linear Regulation

Figure 3–7 shows a discrete zener/transistor linear regulator circuit for high currents. If a very high current must be regulated, and all of this current passes through the series-pass transistor, the transistor may not be able to dissipate all of the heat satisfactorily (even though heat sinks are used). This can result in damage to the transistor. To overcome the problem, several transistors can be connected in parallel with each other. Then the string of transistors is connected in series with the supply output as shown.

Transistors Q_3 through Q_6 are connected in series with the supply output, and in parallel with each other, to act as variable series resistors. Current is divided equally among all four so that each transistor dissipates one-fourth of the total power (and heat).

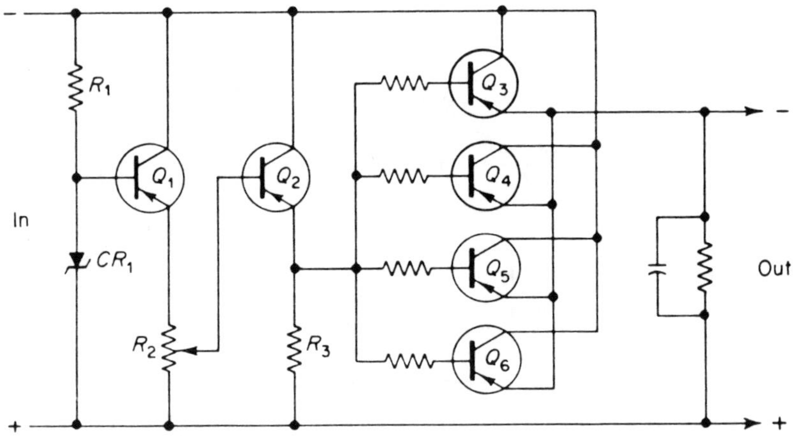

Figure 3–7. Discrete zener/transistor regulator for high currents

The base of Q_1 is held at a fixed voltage (in relation to the emitter) by zener CR_1. However, the voltage across the emitter-collector circuit of Q_1 varies with the supply voltage. If the supply voltage changes for any reason, the voltage across R_2 varies and changes the forward bias on Q_2. In turn, this changes the current through R_3.

Because the voltage across R_3 determines the forward bias on all four transistors (Q_3 through Q_6), the voltage across R_3 determines the amount of resistance offered by transistors Q_3 through Q_6. A change in the Q_3–Q_6 voltage drop changes the supply output voltage to offset the initial change in voltage.

The supply output voltage is set by adjusting R_2, which determines the forward bias on the base of Q_2. The supply is adjusted by connecting a voltmeter across the supply output terminals (usually with the load connected) and setting R_2 for the desired voltage.

3.6 Constant-Current Linear Regulation

Figure 3–8 shows a discrete zener/transistor linear regulator circuit for constant currents. In some applications, it is necessary to produce a constant current from a power supply (rather than a constant voltage).

Notice that a constant-current regulator is sometimes called a CC regulator; this is in contrast to a constant-voltage, or CV, regulator. Most of the regulators described in this book are CV.

In the circuit of Fig. 3–8, transistor Q_1 functions as a variable series resistor in the supply output. Two parallel circuit paths exist: one through zener CR_1 in series with bias resistor R_3, the other through R_1 and Q_1.

If there is any variation in current at the supply output, the current through R_3 varies and produces a change of forward bias on Q_1. In turn, the emitter-collector resistance of Q_1 varies to correct the current flow. The net result is that for every change in R_3 current, there is an equal and opposite change in Q_1 current.

Current output of the circuit is set by R_1. The current remains constant (within limits) despite any load changes. However, the supply output voltage varies with any load changes.

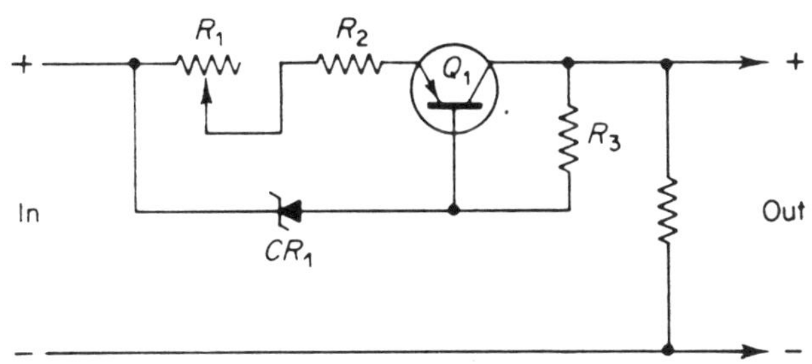

Figure 3–8. Discrete zener/transistor regulator for constant currents

3.7 Overload Protection

Figure 3–9 shows an overload-protection circuit for the series regulator of Section 3.3. Notice that the regulator portion of the circuit is the same as shown in Fig. 3–5.

A series linear regulator is subject to damage caused by overloads. (This is usually not a problem in shunt regulators.) The overload can be in the form of a high input voltage or an excessive output load. Either way, the series-pass transistor can burn out. However, overloads caused by high input voltage are not likely, because any prolonged period of high input voltage usually burns out the basic power-supply components first (transformer, rectifier diodes, etc.). A possible exception is where the series-regulator design is on the borderline of safety.

On the other hand, an excessive output load (low load resistance and high output current) is quite common. (A screwdriver shorted across the regulator output terminals can do the job nicely!) For these reasons, any series regulator that may be subject to overloads should be provided with an overload-protection circuit.

The circuit of Fig. 3–9 is controlled by the voltage across R_3. All of the load current passes through R_3 and develops a corresponding voltage. When the load current is below a certain value (at a safe level below the maximum rating of the series-pass transistor), the voltage drop across R_3 is not sufficient to forward-bias Q_2. So Q_2 remains nonconducting as long as the load current is at a safe level.

If diode CR_2 is silicon, there is about a 0.5-V drop from the output terminal to the emitter of Q_2. Also, if Q_2 is silicon, another 0.5-V drop is required before Q_2 is turned on. As a result, the drop across R_3 must be about 1 V or more (typically 1.1 V) before Q_2 turns on. The value of R_3 is chosen so that the drop is about 1.1 V when the output load current is at the maximum safe level.

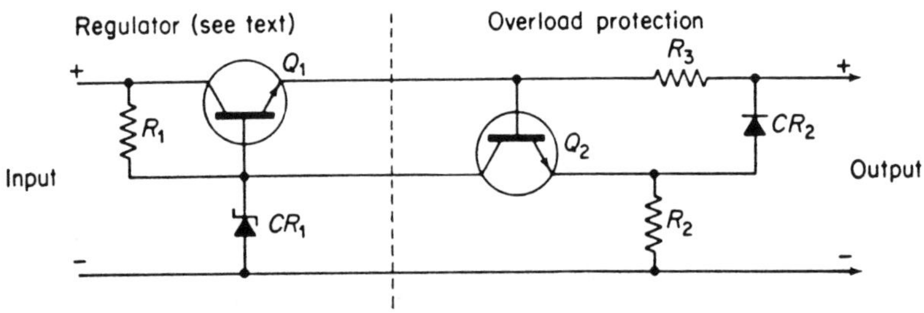

With Q_2 and CR_1 silicon, drop across $R_3 \approx 1.1$ V, at maximum current output
Output voltage ≈ output from regulator − 1 V
Output current and power ≈ same as regulator
$R_2 \approx 10 \times$ load resistance, at maximum current
Load resistance ≈ output voltage / load current
R_3 (in ohms) ≈ 1.1 / maximum load current
R_3 (in watts) ≈ (maximum load current)^2 × R_3 resistance

Figure 3–9. Overload-protection circuit for series regulator

When Q_2 is turned on, part of the current through R_1 is passed through Q_2, drawing current away from the base of Q_1. Transistor Q_1 is cut off (or partially cut off), preventing (or limiting) current flow to the output load. When the load is returned to normal, the drop across R_3 is below 1.1 V, and transistor Q_2 is turned off.

3.7.1 Overload-Protection Design Considerations

The following paragraphs summarize the design considerations for a discrete-transistor linear regulator with overload protection. Notice that the series-pass transistor (Q_1) will probably require a heat sink (see Chapter 2) if the power dissipated by the transistor exceeds 1 W.

3.7.2 Regulator Circuit

Design considerations for the regulator portion of the circuit are the same as described in Sections 3.3.1 through 3.3.7.

3.7.3 Output Voltage

The output voltage from the overload-protection circuit is about 1 V less than the output from the regulator. This is caused by the drop across R_3. The current and power capacity of the overall circuit is about the same as for the regulator circuit.

3.7.4 Transistor Characteristics

Overload-control transistor Q_2 must be capable of withstanding the full output voltage. However, the current (or power) requirements for Q_2 are not high in comparison to those of Q_1. Transistor Q_2 is off and dissipates no power except in case of an overload. When overload occurs, current through Q_2 is limited by emitter resistor R_2.

3.7.5 Emitter Resistor

The first trial value for R_2 should be 10 times the load resistance, at maximum current. For example, if the output voltage is 8 V and the maximum load is 0.4 A, the load resistance is 8/0.4, or 20 ohms. Use 200 ohms for R_2.

The current set by R_2 should be sufficient to drive Q_1 into cutoff at maximum current. For example, assuming a load current of 0.4 A and an output of 8 V, the nominal current through R_1 is 0.4×0.05, or 0.02 A (20 mA).

With the output at 8 V and a 0.5-V drop across CR_2, the drop across R_2 is 7.5 V. With R_2 at 200 ohms and a 7.5-V drop, the current through R_2 (and the emitter-collector of Q_2) is about 37 mA. This is almost twice the nominal 20-mA current through R_1, so Q_1 should be cut off at maximum output current.

3.7.6 Base Resistor

The first trial value of R_3 should be such that the voltage drop is 1.1 V when maximum load current is reached. For example, using the same 0.4-A load, the value of R_3 should be 1.1/0.4, or 2.75 ohms. Use the nearest standard value of 2.7 or 3.0 ohms. The base resistor must be capable of passing the full-load current without overheating. In this example, the power requirement should be $(0.4)^2 \times 3$, or 0.48 W. Use the nearest standard of 0.5 W.

3.7.7 Design Example: Overload Protection

Assume that the circuit of Fig. 3–9 is to provide about 2.5 W of power (8.5 W at 0.3 A). The input voltage varies from 11 to 15 V. A silicon PNP transistor capable of dissipating 5 W is available (with heat sink). The output must be maintained at 8.5 V, at any input voltage between 11 and 15 V, and at any load current up to 0.3 A (300 mA). If the output is shorted or the load resistance drops drastically, the output current must be limited to a safe level.

Verify that transistor Q_1 is capable of handling the full voltage, current, and power. The maximum voltage applied is 15 V, with a maximum current of 0.3 A. This results in a maximum power dissipation of 15×0.3, or 4.5 W. The 5-W transistor should provide a satisfactory margin of safety, because the maximum emitter-collector differential never reaches 15 V (probably never more than 7 or 8 V).

Because the output must be 8.5 V and there is a drop of about 1 V across R_3, the output from Q_1 must be 8.5 + 1, or 9.5 V. Because Q_1 is silicon, the zener CR_1 voltage should be 9.5 + 0.5, or 10 V.

The minimum input voltage should be 10 + 1, or 11 V (or greater). In this case, our low input-voltage limit is on the borderline. A minimum input voltage of 12 V is preferable. If the circuit does not provide proper regulation with the minimum (11-V) input, it may be necessary to raise the minimum input or lower the output capability, depending on other circuit requirements.

As a first trial for base resistor R_1, use a resistance that produces a current at the base equal to 5% of the maximum output current, when the input voltage is minimum, or $(11 - 10)/(0.3 \times 0.05) = 66.6$ ohms. Use the nearest standard of 68 ohms. The power rating for R_1 should be based on maximum current when the input voltage is maximum: $I = (15 - 10)/68 = 73$ mA; $P = (0.073)^2 \times 68 = 0.362$ W. Use the nearest standard of 0.5 or 1.0 W.

The current through R_1 is divided between the base-emitter junction of Q_1 and zener CR_1. Because this current is a maximum of 73 mA, CR_1 should also be capable of 73 mA. The power rating for CR_1 should be the zener voltage (10 V) multiplied by the current, or $10 \times 0.073 = 0.73$ W. Use the nearest standard of 1 W. Notice that CR_1 will probably never pass the full 73 mA, because part of the current passes through the emitter-base junction of Q_1. However, the CR_1 power rating should be selected on the basis of maximum current.

The first trial value for R_2 should be 10 times the load resistance, at maximum current. With an output voltage of 8.5 V and a maximum load of 0.3 A, the load resistance is 8.5/0.3, or about 28 ohms. With a 28-ohm load, the value of R_2 should be 28×10, or 280 ohms. Use the next higher standard of 300 ohms.

With R_2 at 300 ohms, the output at 8.5 V, and an approximate 0.5-V drop across CR_2, the drop across R_2 should be 8 V. This produces a current flow through R_2 and Q_2 of about 27 mA. Because the nominal current through R_1 is calculated at 73 mA, it may be necessary to decrease the value of R_2 so that more current can be drawn by Q_2 (away from the base of Q_1) when Q_2 conducts in response to an overload. If experimental tests prove it necessary to lower the value of R_2, try a value of 100 ohms. This results in an available 80 mA through Q_2.

With a possible 80 mA through Q_2 and 10 V at the collector, the maximum dissipation required for Q_2 is 10×0.08, or 0.8 W. However, because the emitter of Q_2 is at about 8 V, the differential between emitter and collector is 2 V. Q_2 will probably never have to dissipate more than about 0.2 W. However, the power requirement and current rating for Q_2 should be based on the maximum possible current.

The first trial value of R_3 should be such that the voltage drop is 1.1 V with 0.3 A, or 1.1/0.3 = 3.7 ohms. Use the next higher standard of 3.9 ohms. Resistor R_3 must be capable of passing the full 0.3 A without overheating. The power requirement is $(0.3)^2 \times 3.9$, or 0.351 W. Use the next higher standard of 0.5 W.

3.8 Classic Feedback Linear Regulator

Figure 3–10 shows the basic discrete linear regulator with feedback. Such circuits have generally been replaced by IC linear regulators (see Chapter 4) or combination IC and op-amp regulators (see Chapter 5). For this reason, we do not dwell on design considerations. However, the following paragraphs describe operation of the circuit. This provides a basis for understanding the feedback principle found in all linear regulators (IC, op-amp, combination, etc.).

In operation, the output voltage remains within about 0.5 V of the voltage at potentiometer R_3 (at the base of Q_3). Any variations in load (which might result in an undesired change in voltage across the load) cause corresponding changes in current through the base-emitter of Q_3.

Transistors Q_1 and Q_2 are connected as a Darlington pair or compound. Any variation in output current (caused by load changes) changes the base current of Q_2. These changes are amplified through Q_1 and Q_2. The change in emitter-collector current of Q_1 opposes any change in output current. Q_1 is the series-pass transistor. CR_2 provides a constant voltage across R_3, despite any changes in input voltage. As a result, the circuit provides a constant output (in the presence of input-voltage and output-current changes).

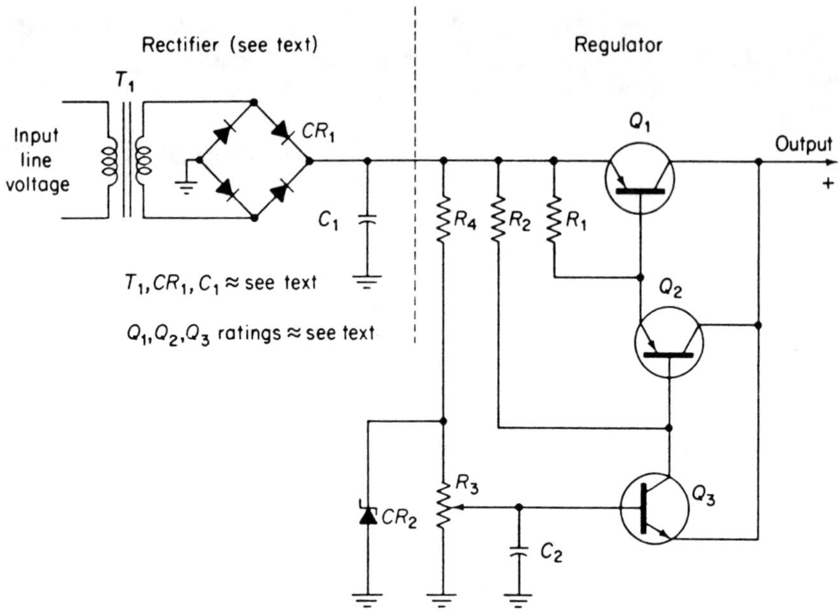

Output voltage ≈ voltage at R_3 (± 0.5 V)

Max output voltage ≈ Zener voltage of CR_2 ≈ minimum output voltage from rectifier − 2 or 3 V

R_3 ≈ max output voltage /(max load current × 0.05)

Current thru R_3 < 5% max load current

R_4 (in ohms) ≈ minimum drop across R_4 /1.2 × current thru R_3

Figure 3–10. Basic discrete linear regulator with feedback

CHAPTER 4

IC Linear-Regulator Basics

Basic IC linear-regulator circuits include IC voltage regulators, current regulators, short-circuit and overload protection, current boosting, current foldback, shutdown, and three-terminal regulators. Worked-out design examples for these circuits, using off-the-shelf ICs, are described in Chapter 7. This chapter concludes with a discussion of IC linear-regulator specifications.

4.1 IC Voltage Regulators

Like discrete voltage regulators, IC voltage regulators provide an output voltage that remains constant (within certain limits) despite changes in load, input voltage, temperature, and so on. In many cases, IC voltage regulators function as the control element in a voltage-regulator circuit (see Chapter 5). However, there are many IC regulators that operate entirely without external components. The following is a brief summary of IC voltage-regulator types available from many manufacturers.

Fixed-output voltage regulators maintain the output of a power supply at a fixed voltage and provide positive and/or negative regulation at various current levels. Fixed-output regulators normally require no external add-on components. However, some manufacturers recommend that input and/or output capacitors be used if the regulator is located some distance from the power-supply filter. Fixed regulators are ideal for on-card regulation of subsystems, affording possible money-saving and performance-improvement advantages in applications where total system regulation is not required.

Variable-output voltage regulators can be tailored to regulate at any specific output voltage (within a given range) through the use of external resistors. A specific output current (at the selected voltage) is available directly from the IC regulator. Increased output current can usually be obtained through the use of external current-boosting circuits.

Current regulators provide a constant current rather than a constant voltage. In many cases, the basic IC voltage regulator can be converted to provide constant-current operation with a few added external components.

Floating regulators for high-voltage operation can deliver hundreds of volts and are limited only by the breakdown voltage of the associated external series-pass transistor.

Dual-tracking regulators provide balanced positive and negative output voltages. Generally, the dual-tracking regulators are set to provide a specific output voltage (such as ±15 V), but an external adjustment can change one or both outputs simultaneously (say, from 8 to 20 V).

Finally, there are *low-temperature-drift regulators.* Such regulators are usually for low-voltage operation, where precision regulation is required.

4.1.1 Basic IC Voltage Regulator Operation

Figure 4–1 shows a combined block diagram and simplified schematic of a basic IC voltage regulator. As shown, the basic regulator is divided into four parts or sections: control, bias, DC-level shift, and output.

The *control section* provides a means of starting and stopping the regulator action. Not all IC regulators have a control section; when it is omitted, the IC is less expensive but somewhat limited in usefulness (in today's sophisticated equipment).

The *bias section* provides a fixed, temperature-compensated reference voltage V_{REF}, typically about 3 or 4 V. Because the desired output voltage is rarely an exact value between 3 and 4 V, it is necessary to "multiply" the reference voltage to some higher exact value.

The *DC-level shift section* performs the necessary "multiplication" of the reference voltage. The available output voltage reference $V_{O(REF)}$ at pins 6 and 9 is determined by the ratio of resistors R_1 and R_2, which set the feedback (and thus the gain) of the reference amplifier. The $V_{O(REF)}$ can be higher or lower than the fixed V_{REF} supplied by the bias section.

The *output section* is essentially a unity-gain, differential amplifier, followed by a pair of NPN transistors. The differential-amplifier inputs consist of $V_{O(REF)}$ at pins 6/9 and the output voltage V_O at pins 1/5. $V_{O(REF)}$ and V_O are the same when the regulator is functioning normally (at the selected voltage).

Any change in output voltage unbalances the differential amplifier and changes the bias of series-pass transistor Q_4. The bias change increases or decreases the current through Q_4 as necessary to offset the initial change in output voltage. The maximum output voltage is determined by the minimum input-output differential (the smallest amount of voltage across Q_4 that allows normal operation).

Note that the differential amplifier has a frequency-compensation capacitor C_C at the collectors. This external capacitor serves the same purpose as the frequency-compensation capacitor in an op-amp. The value of the compensating capacitor is specified on the IC data sheet. Also notice that the series-pass transistor Q_4 has an external capacitor at the output. This capacitor, C_O, is a conventional filter.

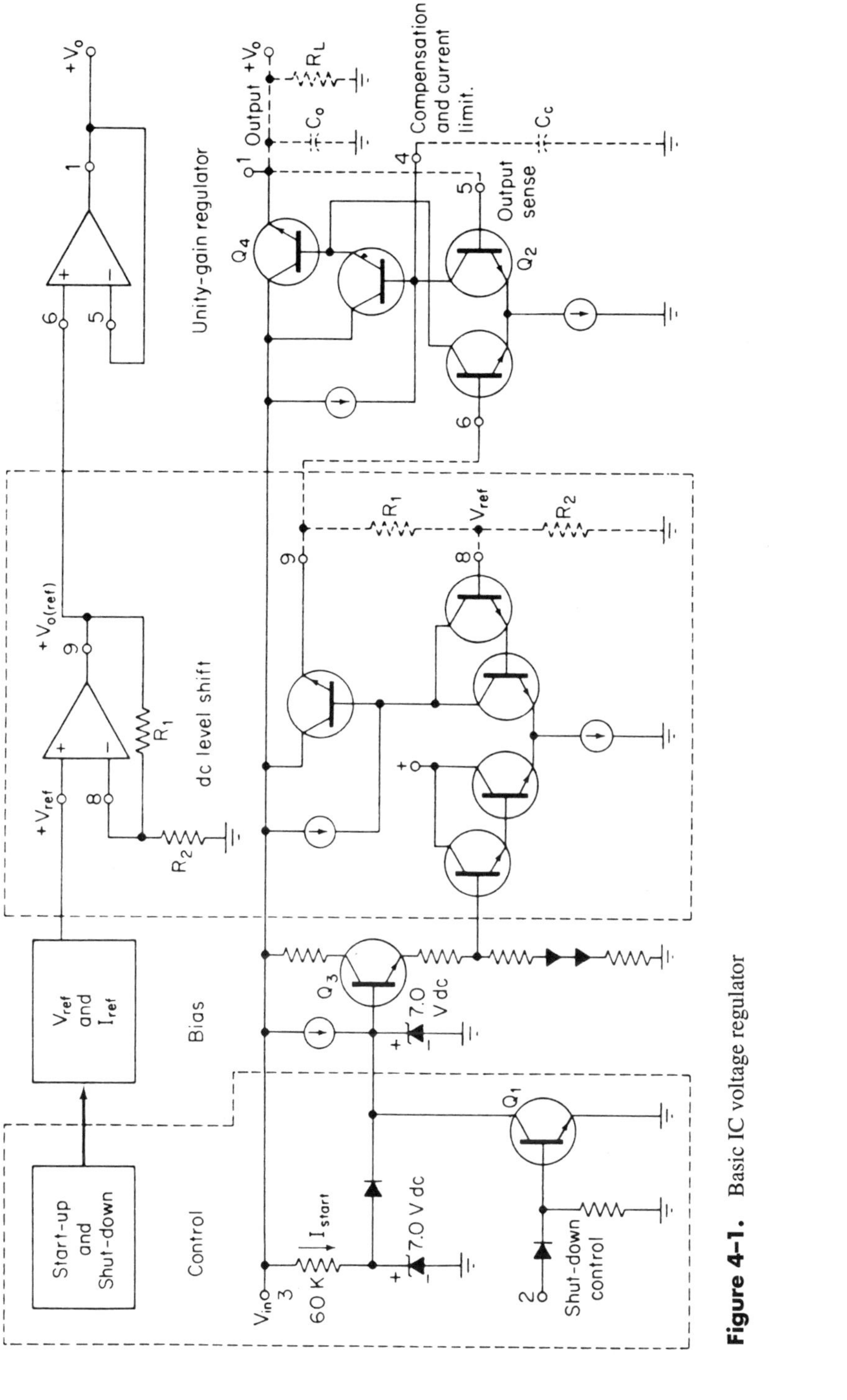

Figure 4–1. Basic IC voltage regulator

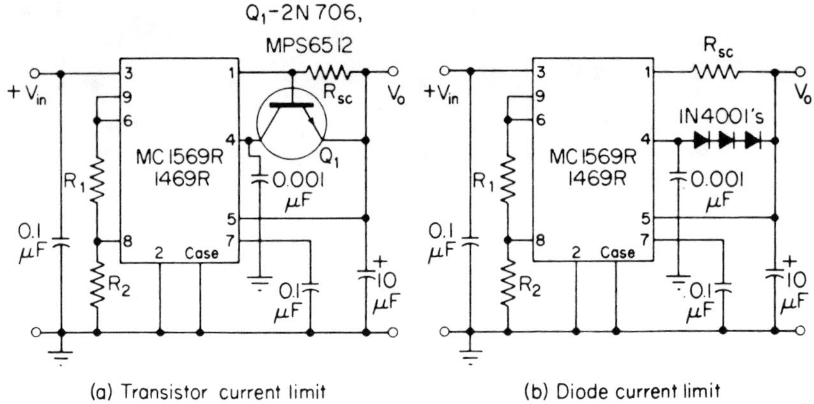

Figure 4–2. Short-circuit and overload-protection circuits

4.2 Short-Circuit and Overload Protection

Figure 4–2 shows typical short-circuit- and overload-protection circuits. Both circuits provide for current limiting and are not to be confused with any thermal-overload circuits found in many IC regulators. The current-limiting circuits shut down the IC when the output current exceeds a desired value. The thermal-overload circuits provide shutdown if the IC junction temperature exceeds a safe level.

With either configuration shown in Fig. 4–2, the transistor (or diode string) is forward-biased when the load current creates a drop across the series resistor R_{SC} equal to the base-emitter drop of Q_1 (or equal to the total diode voltage drops). When R_{SC} is of the proper value, the drop across R_{SC} equals the base-emitter drop (or total diode voltage drops) only if the IC regulator maximum-current limit is reached or approached.

From a user's standpoint, external off-the-chip short-circuit transistors (or diodes) offer some advantages. If either the transistor or diodes are on the chip, the threshold voltage is altered by IC temperature rise. (If the temperature rises, the transistor or diodes pass current at a lower voltage drop.) As more current is drawn, even normal load current, IC temperature can rise quite rapidly (particularly if the IC case is not on a heat sink), and current limiting may set in long before the design value is reached.

No matter what system is used, when the threshold point is reached, the transistor or diodes draw base current away from the NPN transistor pair (at pin 4 of Fig. 4–1) and thus limit output current to a desired maximum.

4.3 Current Boosting

Figure 4–3 shows a typical boosted-current circuit (with current limit) using NPN transistors. Current boosting is often used for higher load current or better regulation.

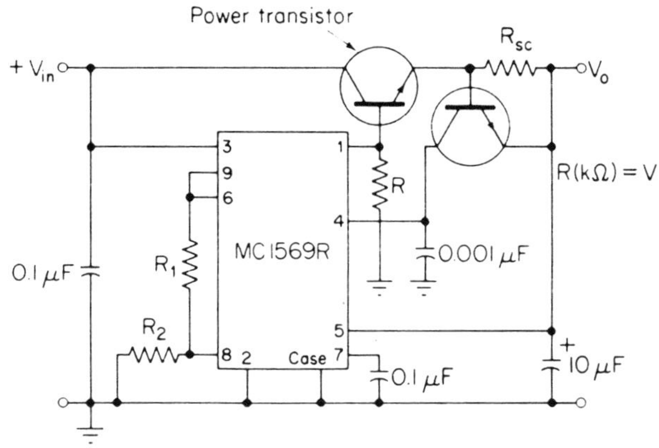

Figure 4–3. Boosted-current circuit with current limit

Even though the IC regulator (Fig. 4–1) may be capable of delivering the desired current, boosting reduces junction-temperature rise, because the series-pass transistor (which dissipates most of the heat) is separate from the IC. So any temperature-related problems (such as reference-voltage drift caused by temperature increases) are minimized.

In the circuit of Fig. 4–3, the load current is passed through the series-pass power transistor. In turn, the series-pass transistor is controlled by the IC regulator output (pin 1). This type of circuit can also be used to increase efficiency at low output voltage because the external series-pass transistor requires only a volt or so to remain active.

4.4 Current Regulators

Figure 4–4 shows the basic IC regulator (Fig. 4–1) used as a current regulator, both with and without current boosting. In both configurations the output current passes through the load R_L, and through resistor R_1. The voltage drop across R_1 is returned to the IC regulator at pin 5. This voltage is compared with the reference voltage (at pin 6) set by the R_2/R_5 ratio. Any deviations in current are amplified and corrected to maintain load current constant to within less than 1% (typically 0.05%).

4.5 Shutdown Circuits

Many linear IC regulators have some means for shutdown and startup. In the circuit of Fig. 4–1, the IC is controlled by transistor Q_1 and the associated circuit. In normal operation (when the IC regulator is functioning), the shutdown-control input at pin 2 is open or connected to a voltage near ground. In this condition Q_1 is nonconducting and has no effect on normal operation of the IC.

Transistor Q_1 shunts the internal 7-V reference zener. Should Q_1 be turned on and saturated, the zener is shorted to ground. Because all internal current sources are

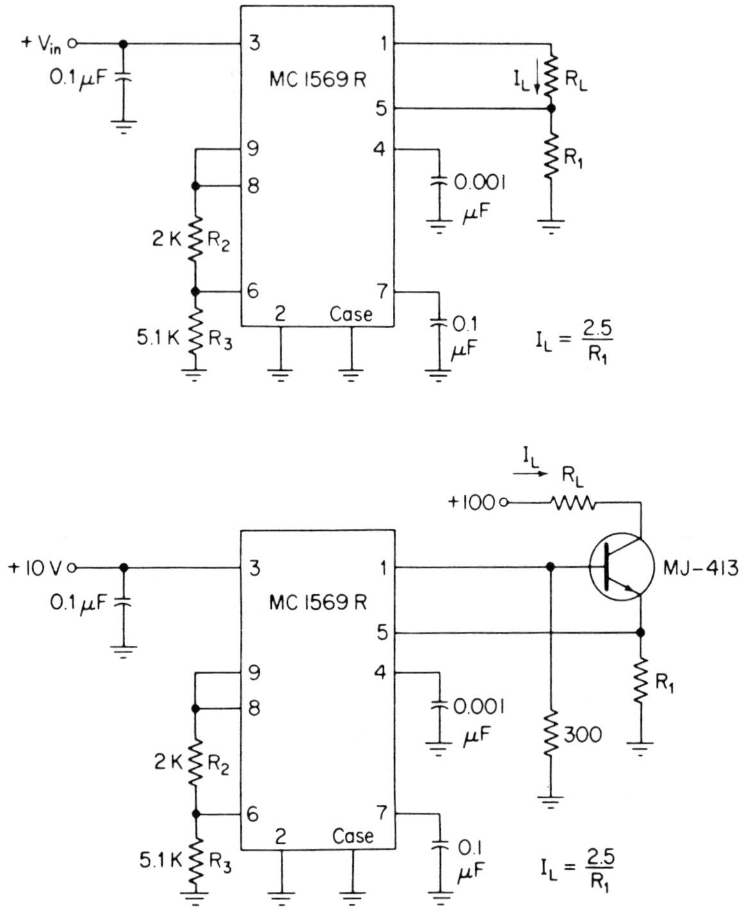

Figure 4–4. Basic IC voltage regulators used as current regulators

biased to the internal reference, the sources cannot function, and the IC regulator stops operation (series-pass transistor Q_4 is shut off).

The IC is shut down by applying a voltage at pin 2. The voltage must be large enough to drive Q_1 into saturation. The IC returns to normal operation once the voltage is removed. The following paragraphs describe the external circuits required to shut down the IC in a variety of operating configurations.

4.5.1 Logic-Signal Shutdown

Figure 4–5 shows how shutdown of the IC regulator can be controlled by a logic gate. It is assumed that the IC regulator is operating in the normal mode (as a positive regulator, referenced to ground) and that the logic gate is of the saturating type, operating from a positive supply. The gate can be TTL (or even the ancient DTL/RTL) as long as the gate output uses an active pull-up resistor.

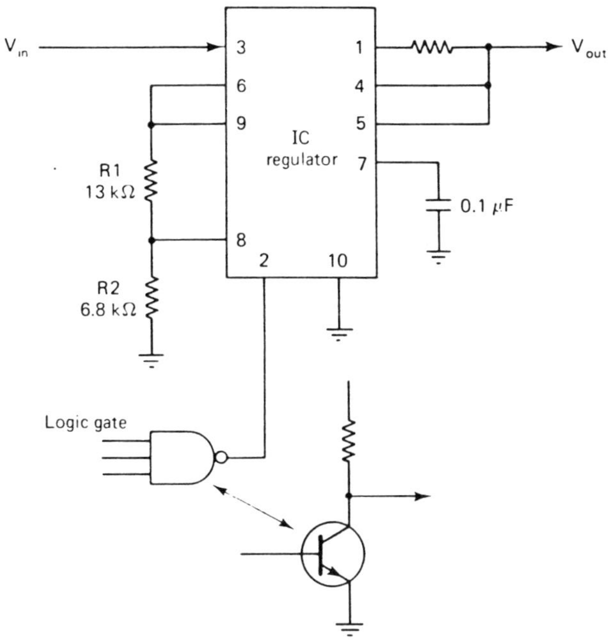

Figure 4–5. Shutdown of IC regulator with logic gate

A typical application for the circuit of Fig. 4–5 is for remote systems where power is to be conserved until a particular subsystem is needed to function. At such time, the IC regulator is turned on via a control signal (the voltage at pin 2 is removed or goes low), the regulator performs the required task, and the subsystem is then shut down (the voltage at pin 2 goes high).

4.5.2 Temperature Control or Shutdown

Figure 4–6 shows how shutdown of the IC regulator can be controlled (externally) by temperature. This is not to be confused with internal thermal-overload circuits found in many IC regulators.

Unlike discrete-component linear regulators (see Chapter 3), most of an IC regulator chip is occupied by the series-pass transistor (because this transistor handles most of the current). The temperature of the entire IC is set by the series-pass transistor. Variations in load cause the transistor temperature to change, thus changing the temperature of the remaining components on the IC. This "thermal feedback" or "temperature feedback" can cause thermal runaway: An increase in temperature causes an increase in current, which in turn causes a further increase in temperature, and so on, until the IC burns out.

It is possible to use thermal feedback to control the IC (externally) and to initiate shutdown if an unsafe temperature is reached. Two methods are shown in Fig. 4–6. One method (a) uses junction temperature, or T_J; the other method (b) uses ambient temperature, or T_A.

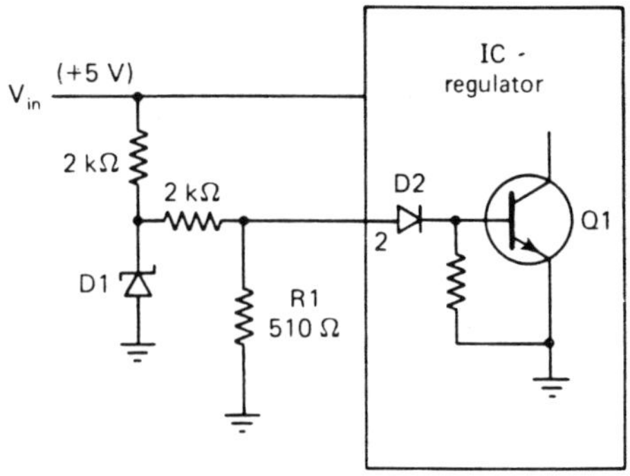

(a) Junction temperature T_J

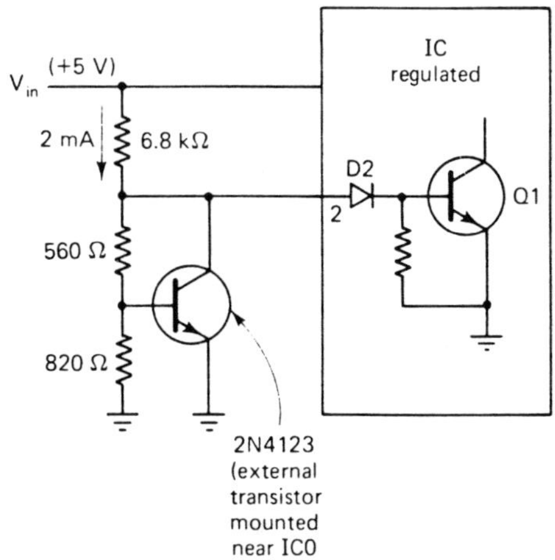

(b) Ambient temperature T_A

Figure 4–6. Shutdown of IC regulator with temperature control

When *junction temperature* T_J is used, the temperature of the D_2 and Q_1 junctions can be considered the same as that of the series-pass transistor. As temperature increases, the voltage required to turn on D_2 and Q_1 (and thus turn off the regulator) drops. Thus, if a fixed voltage is applied, D_2 and Q_1 turn on (the regulator turns off) when a predetermined temperature is reached. The external zener D_1 provides the fixed voltage (which can be set by varying R_1). Zener D_1 is usually mounted *away from* the IC so that heat from the IC does not affect D_1.

When *ambient temperature* T_A is used, the fixed reference for control is set by the drop across the junction of an external transistor (a 2N4123 in this case). Because the voltage drop is controlled by ambient temperature, the control voltage (and turn-on/turn-off point) is set by the ambient temperature. The external transistor is usually mounted *near* the IC so that a large increase in IC heat changes the ambient temperature around the transistor.

4.5.3 Combined Temperature and Short-Circuit Shutdown

The temperature-control or shutdown systems described here are often used in conjunction with the short-circuit and overload circuits described in Section 4.2. When a short circuit or overload occurs, increased current is drawn through the IC components and the IC can heat up, possibly to a dangerous level. Using both overload and temperature, the IC turns off when the temperature reaches a predetermined point during overload. When the IC is cool enough, the regulator returns to the on condition. If the short or overload still exists, the heat process begins again, and the IC continues turning on and off as long as the output is shorted.

The technique of combining both temperature and short-circuit protection provides control of both the maximum short-circuit current and the maximum junction temperature. Such complete protection is often considered superior to the current-foldback system described in Section 4.6.

4.5.4 Current-Boost Shutdown

Figure 4–7 shows a method of temperature-controlled shutdown used when the IC regulator is operating in a current-boost circuit, as described in Section 4.3.

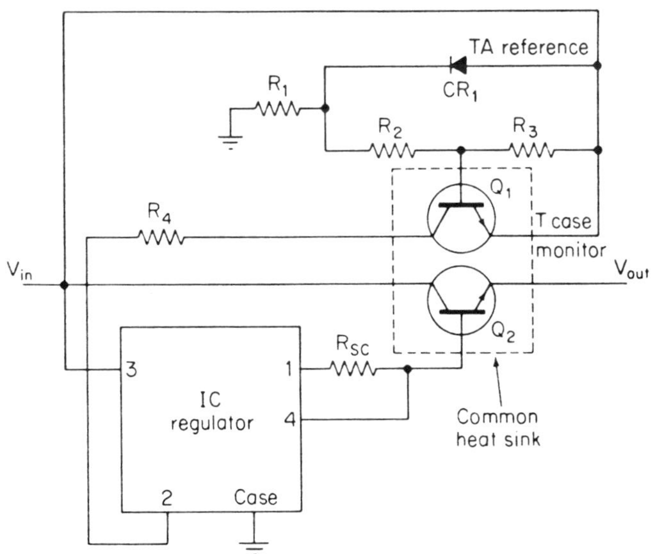

Figure 4–7. Temperature-controlled shutdown with IC regulators used in current-boost circuit

The external series-pass transistor (being controlled by the IC regulator) is mounted on a common heat sink with a monitor transistor.

In this configuration, the external series-pass transistor Q_1 heats when the output is shorted or there is an overload, while the monitor transistor Q_2 senses the heat-sink temperature. The monitor transistor conducts when the heat-sink temperature exceeds the normal ambient temperature by a predetermined amount. This applies a signal to pin 2 and shuts off the regulator. The point at which the monitor transistor conducts is set by the values of R_2/R_3. Resistor R_4 limits the drive into pin 2.

4.5.5 Output Short-Circuit Shutdown

Figure 4–8 shows an IC-regulator shutdown circuit using an output short circuit for control. External transistor Q_1 is normally saturated because of the base drive supplied through resistor divider R_1 and R_2. With Q_1 in saturation, the voltage on pin 2 of the IC is below the threshold of the IC control circuit.

If the IC output is short-circuited, Q_1 turns off, the voltage at pin 2 rises, and the regulator shuts down. The values of R_1 and R_2 are selected to saturate Q_1, as well as to sink minimum regulator load current (typically about 1 mA). Capacitor C_1 provides an RC time constant that prevents shutdown when V_{IN} is initially applied (before Q_1 can saturate). When the output short circuit is removed, the regulator must be manually reset before full regulation is restored.

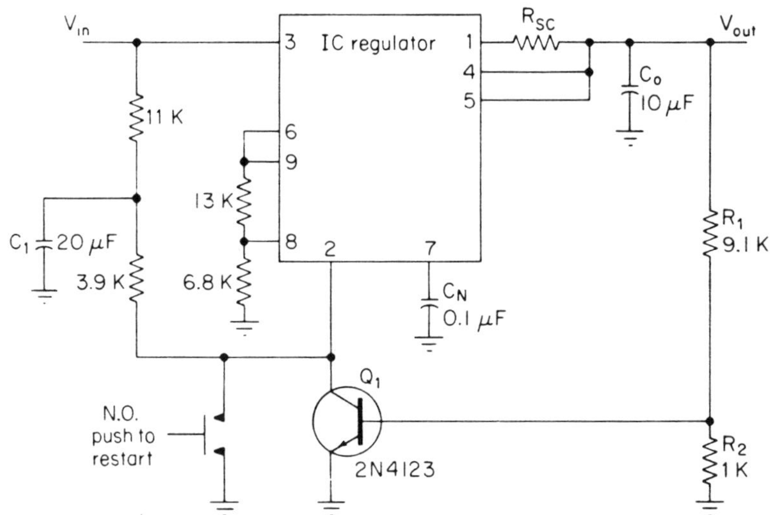

Figure 4–8. IC regulator shutdown using output short circuit

4.6 Current Foldback

Figure 4–9 shows a short-circuit shutdown circuit for IC regulators without a control section. This circuit reduces the regulator output voltage and current to zero (or near zero) when the load current reaches a certain limit. The technique is called *current foldback* because both the voltage and current drop back (or fold back) to some low (safe) level when the output current reaches a certain limit (as shown by the graph in Fig. 4–9).

The voltage drop across R_2 is set to about twice the base-emitter drop of Q_1 (somewhere between 1 and 1.4 V) when the output voltage is normal. For Q_1 to conduct, the voltage across R_3 (that senses the output load) must be about equal to the drop across R_1, plus the base-emitter voltage. By proper selection of resistance values, this occurs only if the current limit is reached. When Q_1 conducts, the collector connected to pin 4 diverts current drive from the series-pass transistor (Q_4 in Fig. 4–1), producing current foldback.

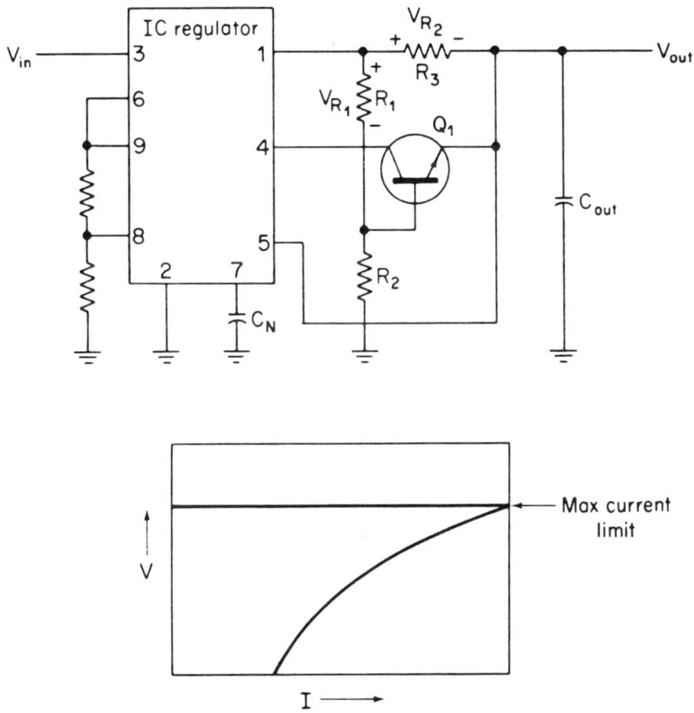

Figure 4–9. Short-circuit shutdown for IC regulators without a control section (current foldback)

4.7 Three-Terminal Regulators

The three-terminal regulator is the simplest form of linear IC voltage regulator. As the name implies, the IC has only three active terminals: input, output, and common or ground.

The input terminal is connected to an unregulated voltage source (typically at the output terminal of a rectifier package), and the common terminal is connected to ground (or to the negative terminal of the rectifier package). The voltage at the output terminal (connected to the load) remains constant despite variations in load current (within limits).

The data sheets for three-terminal regulators specify input/output voltage and current limits and show what external components (if any) are required. We do not duplicate this data-sheet information here. Instead, we show some classic linear power-supply circuits using three-terminal regulators. (Chapter 7 has several worked-out design examples using three-terminal regulators, including how the regulator range can be extended and made adjustable.)

As shown in Figs. 4–10 and 4–11, the three-terminal regulator can be used directly to produce a constant voltage (such as the +24-V and +33-V supplies) or used as the control element for a series-pass transistor (such as the 5-V supplies). The component types and values shown in Figs. 4–10 and 4–11 can be used as first trial values for these classic linear power supplies.

4.8 Linear IC-Regulator Specifications

Unfortunately, the specifications for linear IC regulators of all manufacturers are not consistent. The following paragraphs briefly explain the terms used in most data sheets. If linear-supply designers fully understand these terms, they should have no difficulty in specifying a linear IC regulator or understanding the data sheets of any manufacturer.

4.8.1 Output Voltage Range

This is the range of output voltages over which the specifications apply. In many IC regulators, the output voltage is established by a user-selected external resistive-divider network (such as R_1 and R_2 in Fig. 4–1).

4.8.2 Output Current

All IC regulators are capable of supplying a certain amount of current to a load without the use of external transistors. The magnitude of this current is specified on the data sheet as a minimum and a maximum. For example, a particular IC regulator might be capable of supplying 500 mA for any output voltage from 2.5 to 37 V. However, the same regulator must supply at least 1 mA to maintain full regu-

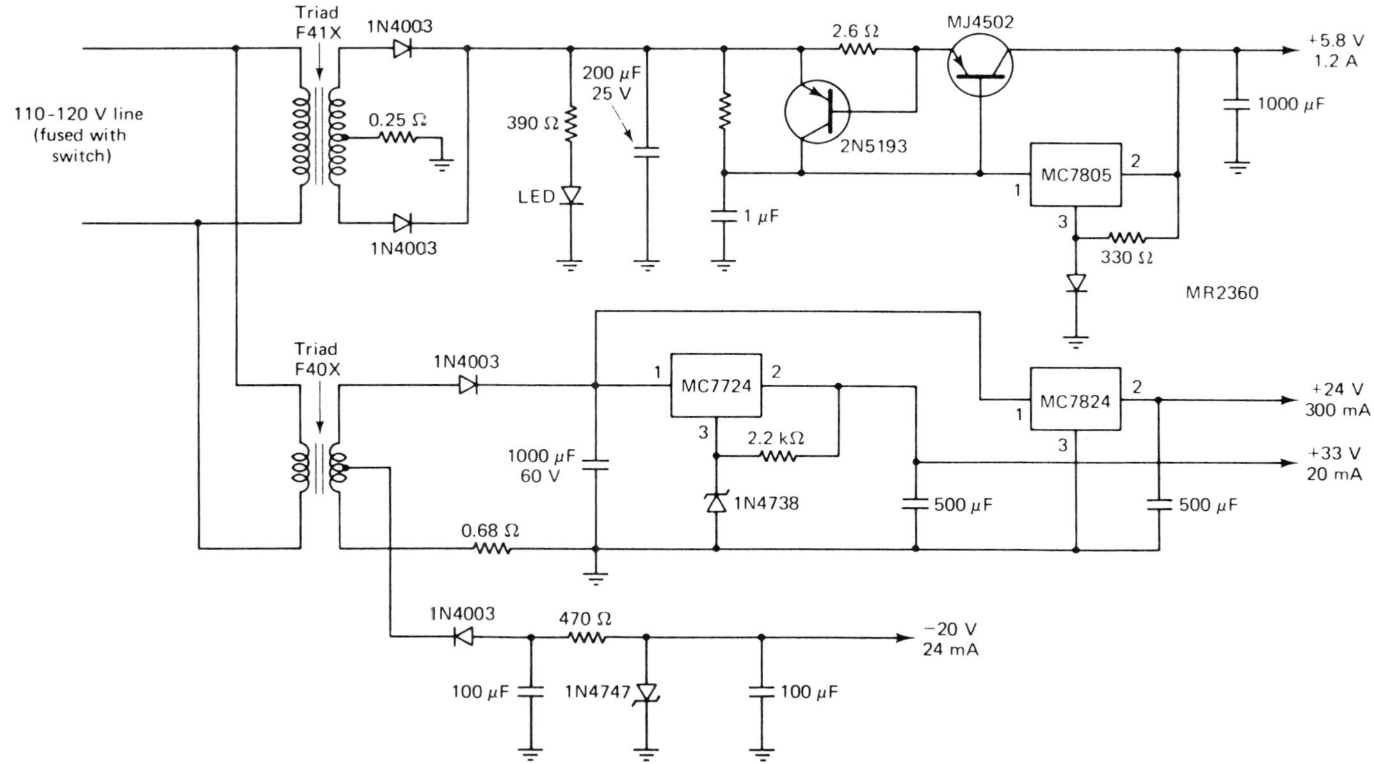

Figure 4–10. Typical linear supply with three-terminal regulators

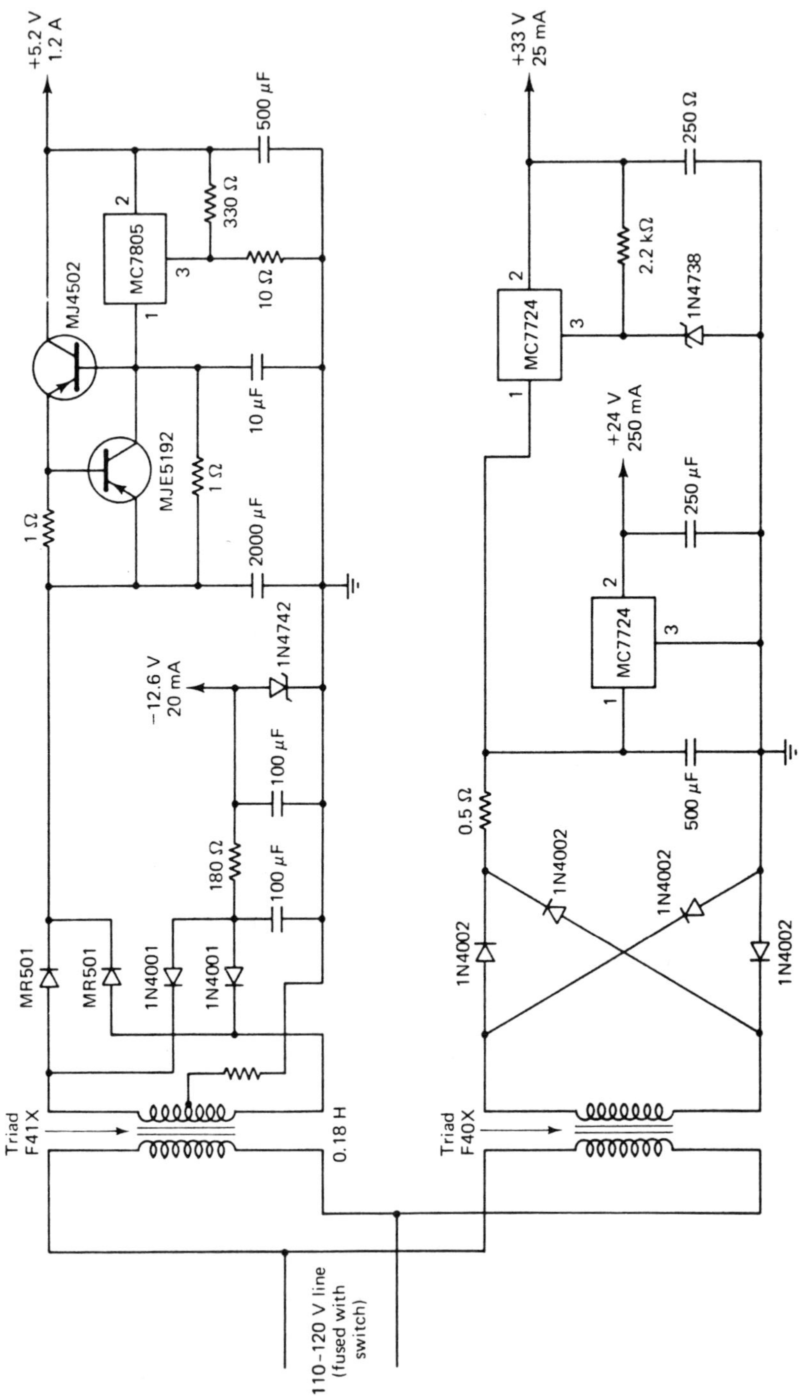

Figure 4–11. Alternate linear supply with three-terminal regulators

lator action. For this reason, the IC data sheets often specify that a resistance be placed across the output to draw a minimum of 1 mA at the selected voltage.

4.8.3 Input-Output Voltage Differential

The input-output voltage differential ($V_{IN} - V_O$) is the voltage required to bias the circuit that supplies drive current to the series-pass transistor. As the input voltage drops and approaches the output-voltage level, the regulator *drops out of regulation* at the minimum value of this specification. The problem of regulator dropout is especially critical in low-voltage circuits such as those found in battery-operated equipment. Dropout problems are discussed throughout Chapter 7, particularly in Section 7.11.

4.8.4 Input or Supply Voltage Range

The input or supply voltage range is specified on the data sheet as a minimum and a maximum. This value indicates that a certain minimum voltage must be present to properly bias the zener (or other reference) on the IC. This voltage must also exceed the output voltage by at least the specified input-output voltage differential. For example, if the $V_{IN} - V_O$ is 2-V minimum and 3-V maximum and the desired output is 10 V, the input or supply voltage must be at least 12 V, and possibly 13 V.

4.8.5 Load Regulation

Load regulation is the percentage change in output voltage for a steady-state change in load current from the minimum to maximum value specified. This is expressed as follows: load regulation = $(V_{NL} - V_{FL})/V_{NL} \times 100$ ($\%V_O$), where V_{NL} is the output voltage with minimum load current, V_{FL} is the output voltage when the full-load current is being drawn, and the units are a percentage of V_O ($\%V_O$).

Notice that some data sheets specify load regulation directly in terms of output-voltage change for minimum-to-maximum load change. For example, one of the regulators described in Chapter 7 describes load regulation as a typical 5-mV change (and a maximum 30-mV change) when the load changes from 1 mA to 100 mA, when the typical output voltage is 15 V.

4.8.6 Output Impedance

Output impedance Z_O is a small-signal parameter. Z_O indicates the ability of the regulator to prevent power-supply voltage changes created by fluctuation currents drawn at signal frequencies by circuit load. High values of Z_O can create undesirable coupling between circuits powered by a common regulator, possibly resulting in oscillation. Typical output impedances are less than 1 ohm.

4.8.7 Line or Input Regulation

Line or input regulation is the percentage change in output voltage per volt change of the input voltage. The term *line regulation* is used to show the dependence of the output voltage on the power-line variations. Other data sheets list *input regulation* because IC regulators do not operate directly from the power line.

No matter which term is used, the specification can be expressed as a percentage: line or input regulation = $(\Delta V_O/V_O \Delta V_{IN}) \times 100$ (%/Vin). In this case, ΔV_O is the change in output voltage V_O for the input change ΔV_{IN}, and %/V_{IN} are the units.

Notice that some data sheets specify line or input regulation directly in terms of output-voltage change for a given input-voltage change. For example, a regulator in Chapter 7 describes line regulation as a typical 2-mV change (and a maximum 20-mV change) when the input voltage (output of the power supply) at the regulator changes from 18 to 30 V.

4.8.8 Temperature Coefficient or Drift

The temperature coefficient (or tempco) is the stability of the output voltage over a change in operating temperature. The specification is also listed as *output voltage drift with temperature* on some data sheets and is given in terms of %/°C. For example, a regulator in Chapter 7 describes output voltage drift as a typical 0.005% (and a maximum 0.015%) per °C.

4.8.9 Supply or Quiescent Current

This specification represents the amount of current drawn by the regulator when load current is zero and supply voltage is maximum. For example, a regulator in Chapter 7 describes supply current as a typical 1.5 mA (and a maximum 4 mA) when the input voltage (output of power supply) at the regulator is 30 V and the load current (at the regulator output) is zero.

4.8.10 Ripple Rejection

This specification represents the amount of rejection (or attenuation) to ripple at the regulator input and is expressed in decibels (dB). Generally, the ripple frequency is considered to be 120 Hz, or twice the line frequency of 60 Hz, because the power supply is full wave (see Chapter 1).

Ripple rejection is usually specified at some particular junction temperature, or T_J, because the rejection changes with temperature. For example, a regulator in Chapter 7 lists ripple rejection as a typical 75 dB with a frequency of 120 Hz and a T_J of 25°C. This means that any ripple voltage at the input to the regulator will be reduced by 75 dB at the output.

4.8.11 Short-Circuit Current

This specification represents the amount of current drawn by the regulator when operating into a short circuit. A regulator in Chapter 7 lists short-circuit current as a typical 200 mA with a T_J of 25°C.

4.8.12 Output Voltage Noise

This specification represents the noise voltage generated by the regulator during normal operation. The value is listed in terms of μV over a given frequency range and at a given T_J. For example, a regulator in Chapter 7 lists output voltage noise as a typical 60 μV RMS, over a frequency range from 100 Hz to 10 kHz, at a T_J of 25°C.

4.8.13 Internal Thermal Shutdown

This specification represents the temperature at which the internal-shutdown circuits operate to turn off the regulator. Typically, the temperature is 175°C. As discussed in Chapter 7, thermal shutdown usually starts just before the specified temperature is reached. Also, some designs use external components to shut down the regulator long before the specified temperature is reached. The internal thermal-shutdown circuits are generally regarded as an ultimate fail-safe to prevent the IC from being destroyed by excessive temperatures.

CHAPTER 5

Op-Amp Linear Regulator Basics

This chapter is devoted to IC operational amplifiers (op-amps) used as linear voltage regulators, which are not to be confused with IC packages that function as complete regulator circuits (see Chapter 4). Here the discussion is limited to the use of an IC op-amp as the active or gain element in a linear voltage-regulator circuit. As observed in Chapter 7, the present trend is to combine the gain of an op-amp with the stability of an IC regulator to form a precision linear regulator.

Linear regulators using IC op-amps as the gain elements usually show much better regulation than the packaged IC regulators. This is because of the higher available feedback-loop gain in the op-amp. Voltage drops of less than 0.01% over the entire load range are commonplace, and (with care) 0.001% or better is possible using op-amps.

5.1 Op-Amp Linear Regulator Theory

Figure 5–1 shows the theoretical feedback op-amp form of voltage regulator. The output voltage and reference voltage for this theoretical system can be calculated from the equations shown. However, these equations do not take into account the output impedance of the op-amp or the effects of the load.

The circuit shown in Fig. 5–2 is somewhat more practical. Here the op-amp is followed by an emitter-follower stage that provides the necessary current gain (for a useful output current), as well as low output impedance.

As shown by the equation in Fig. 5–2, the output impedance of the regulator (Z_O) is about equal to the impedance seen at the base of Q_1, divided by the beta of Q_1. In turn, the base impedance is essentially the open-loop output impedance of the op-amp. Thus, the open-loop output impedance of the regulator is equal to the op-amp output impedance divided by the beta of Q_1. When feedback is added, the regulator output impedance drops (by an amount proportional to open-loop gain), as shown by the closed-loop-gain equation.

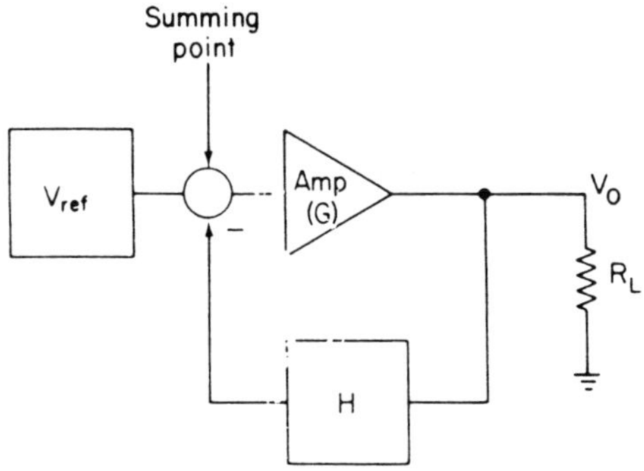

$$V_O = V_{ref}\,\frac{G}{1+GH}$$

G = Amplifier gain (A_{vol})
H = Fraction of V_O fed back to summing point

$$V_{ref} - V_O = V_{ref}\,\frac{1}{1+G}\quad \text{for } H=1$$

Figure 5–1. Theoretical op-amp regulator

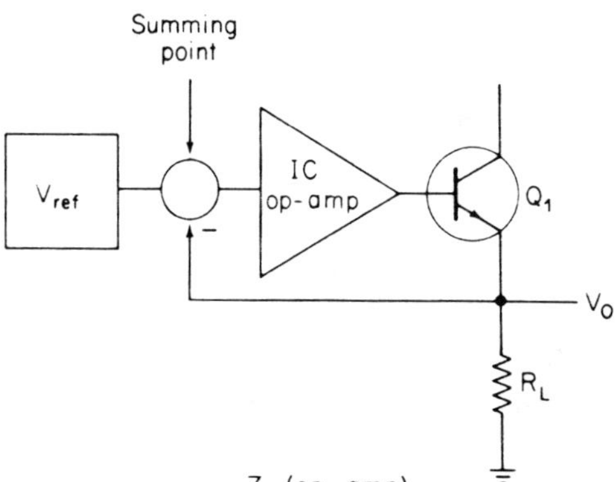

$$Z_O = (\text{Open loop}) \approx \frac{Z_O\,(\text{op-amp})}{\text{Beta } Q_1}$$

$$Z_O = (\text{Closed loop}) \approx \frac{Z_O\,(\text{op-amp})}{(1+G)}$$

Figure 5–2. Theoretical op-amp regulator with improved current capability

Using either circuit (Fig. 5–1 or 5–2), any variations in output voltage (because of changes in load) or in input voltage (changes in V_{REF}) cause a corresponding change in feedback voltage. In turn, this causes the amplifier gain to change in a direction to oppose the initial change.

For example, if V_{REF} is increased, V_O starts to increase, as does feedback. The increase in negative feedback offsets the initial increase in V_{REF}, so the net effect is that V_O remains constant.

In a practical circuit, V_O is never constant. There is always some variation in output, which is expressed as a percentage of regulation. As the discussion in Section 5.2 shows, the percentage of regulation decreases (i.e., regulation improves) when the closed-loop output impedance of the circuit decreases (and when op-amp gain increases).

5.2 Calculating Percentage of Regulation

The circuit of Fig. 5–3 can be used to calculate the *theoretical* percentage of regulation for any feedback regulator. Figure 5–3 shows the regulator as a voltage source V′, with a finite output impedance R_O. One equation for the percentage of regulation involves the voltage source V′ and the output voltage V_O. The alternate equation involves source V′ and the term $I_L R_O$, which is the load current through the regulator output impedance. The $I_L R_O$ term is equal to the voltage drop across R_O, because E = IR. The term $I_L R_O$ also equals the difference between V′ and V_O (V′ – V_O).

$$\frac{1000}{20\,(1 + 100{,}000)} \approx 0.5\ \text{m}\Omega$$

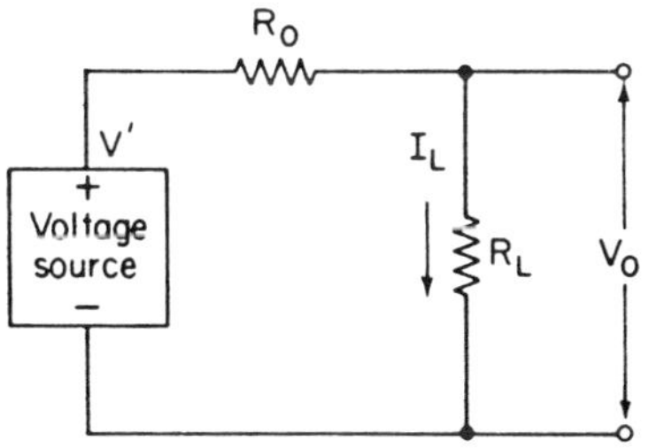

$$V' = \frac{V_{ref}\,G}{1 + GH} \qquad R_O = \frac{Z_O\ (\text{op-amp})}{\text{Beta } Q_1\ (1+G)}$$

$$V_O = V' - I_L R_O$$

$$\text{Percent regulation} = \frac{V' - V_O}{V'} \times 100 = \frac{I_L R_O}{V'} \times 100$$

Figure 5–3. Theoretical percentage of regulation for a feedback regulator

No matter which equation is used, the percentage of regulation is decreased (i.e., regulation is improved) when R_O is decreased. A decrease in R_O indicates a lower regulator-system output impedance, a lower I_LR_O term, a lower voltage drop across R_O, and a smaller difference between V′ and V_O, all of which lower the percentage of regulation.

Using the theoretical circuit of Fig. 5–2, assume that the IC op-amp has an open-loop gain of 100,000 and an output impedance of 1000 ohms, that the emitter-follower has a beta of 20, and that V_{REF} is 10 V. Using the equations of Fig. 5–3, the value of R_O is found as follows: 100/20(1 + 100,000) = 0.5 milliohm.

Assuming a load current I_L of 100 mA and using the I_LR_O equation of Fig. 5–3, the percentage of regulation is (100 mA × 0.5)/10 V × 100 = 0.0005%.

Using the $V' - V_O$ equation of Fig. 5–3, find the same percentage of regulation:

$$V_O = V' - I_LR_O = 10\text{ V} - 0.00005 = 9.99995.$$

$$(V' - V_O)/V' = (10\text{ V} - 9.99995)/10\text{ V} \times 100 = 0.0005\%.$$

From a user's standpoint, a 0.0005% value for regulation is not practical. There are several factors that prevent such a low percentage. The following paragraphs describe the major problems found during practical design.

5.3 Effects of Offset Voltage

Using the theoretical values of the preceding example, the output voltage is supposed to follow the input or reference within about 500 μV. This value can easily be masked by the input offset voltage (V_{IO}) of a typical op-amp.

Unless some provisions are made to neutralize the offset, the output voltage is always removed from the input voltage by an amount equal to V_{IO}. For example, assume that the uncorrected V_{IO} is 1 mV and that the same reference (10 V) as the preceding example is used to find percentage of regulation: (10 V – 9.999 V)/10 V × 100 = 0.01%.

Another effect of V_{IO} is *output drift* with temperature and loading. Assume a V_{IO} drift with temperature of 5 to 20 μV/°C. This shows up not only as added shift between output voltage and reference voltage over a temperature range, but also with variations in load as well.

For example, assume that the temperature increases 50°C for a given load change. If the drift is 20 μV/°C, V_{IO} changes by a value of 1 mV (50 × 20 μV). Even if V_{IO} is neutralized to zero at the lower temperature, the output is offset by 1 mV from the input, resulting in a 0.01% regulation (assuming the values of the preceding example).

In a practical IC op-amp regulator circuit, the output voltage appears to drift after the load is applied or the load is changed. Then, after some time, the output voltage settles to a constant value.

The regulation figures using the equations of Fig. 5–3 are of little value if the V_{IO} drift completely covers the short-term regulation. A typical IC op-amp regulator requires about 1 to 2 minutes to settle after a change in temperature. This can easily interfere with good regulation for rapid load changes.

5.4 Effects of Common-Mode Rejection and Power-Supply Sensitivity

If an IC op-amp receives power directly from the regulator-circuit input voltage as shown in Fig. 5–4, any variation in input voltage can produce some change in input offset voltage V_{IO}. The amount of V_{IO} change for a given power-supply change is a function of op-amp power-supply sensitivity.

The percentage of regulation can be no better than the op-amp power-supply sensitivity. For example, assuming the same 10-V reference, a change in V_{IO} of 1 mV (because of power-supply variation) produces 0.01% of regulation.

Another effect of power-supply variation is that a common-mode signal or error voltage (expressed in dB) is generated when V_{REF} remains constant (as it should) but the power-supply voltage changes. When V_{REF} is made the same as the op-amp power supply, the common-mode error signal is equal to the difference between V_{REF} and the supply voltages. As a guide, any IC op-amp used as the active element in a regulator should have a common-mode rejection ratio of at least 90 dB and preferably 100 dB.

5.5 Effects of Voltage Reference

No matter what regulator circuit is used, output voltage stability over a period of time, and with temperature variations, depends for the most part on the quality of the voltage reference. An IC op-amp, even with very low temperature-drift characteristics, cannot guarantee stable regulation. Considerable care must be used to select a good voltage reference. Here are some points to consider.

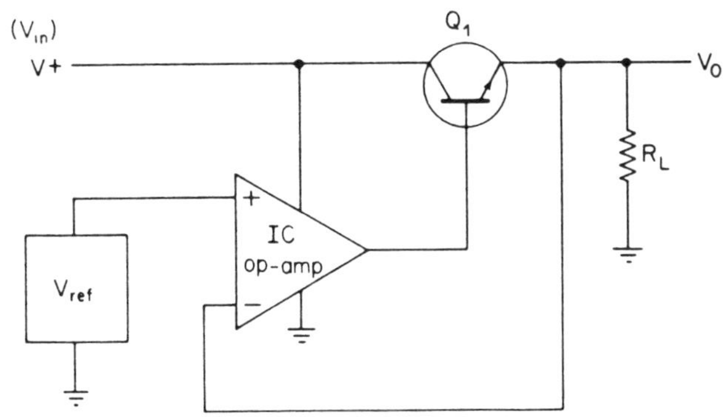

Figure 5–4. IC op-amp regulator powered by line voltage

5.5.1 Zener Voltage Reference

The most obvious solution is to use a zener that has a low temperature coefficient (or a zener that has been compensated by adding forward diodes or temperature-stabilized). Unfortunately, zeners have certain problems.

For example, low-voltage zeners are difficult to use when the regulator must provide a high-voltage output. High-voltage compensated zeners are expensive. Also, when the reference voltage is equal to the output voltage, current from the input voltage must be used to operate the zener. This can cause some line ripple to feed through to the output.

The circuit of Fig. 5–5 is generally satisfactory if a zener is used as the reference for an IC op-amp regulator. The drop across series resistance R_S is equal to the input voltage less the zener voltage (which becomes V_{REF}). Notice that Fig. 5–5 includes equations for trial values of the reference circuit. The following summarizes the use of these equations.

The zener voltage should be about 0.7 times the lowest input voltage. For example, if the input voltage varies between 10 and 15 V, the zener voltage should be 10×0.7, or 7 V. If the design problem is stated in reverse, the minimum input voltage must be 1.4 times the desired zener voltage. Thus, if the required zener (or V_{REF})

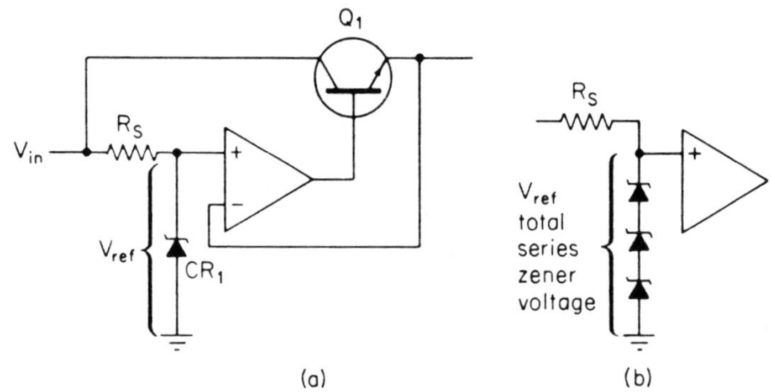

V_{ref} = Zener voltage

CR_1 power dissipation = Zener V x Zener I

Zener V ≈ 0.7 x lowest V_{in}

Min. $V_{in} \geqq 1.4$ x Zener V

Safe Zener power ≈ 3 x load power

$$R_S\text{(in ohms)}\ \frac{(\text{Max. } V_{in} - \text{Zener V})^2}{\text{Safe power dissipation}}$$

V_{in} variation $\leqq$ 30% of max. V_{in}

Load current = Zener V/ load resistance

Load power = Zener V x load current

Safe Zener current ≈ 3 x load current

Figure 5–5. Op-amp regulator powered by linc voltage

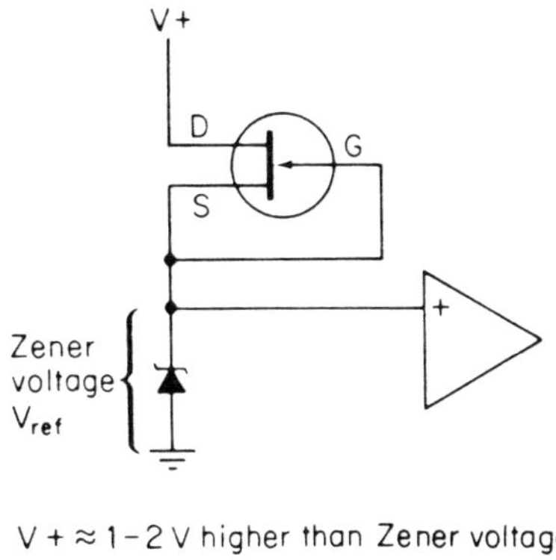

Figure 5–6. Zener voltage reference using FET in place of a series resistance

voltage is 10 V, the minimum input voltage from the basic power supply must be 10 × 1.4, or 14 V.

Because the zener voltage must be about 30% below the input voltage (to ensure that the zener goes into avalanche condition), there is a corresponding drop across the series-pass transistor Q_1 when the zener is used as V_{REF} (and the regulator output must be equal to V_{REF}).

A large drop across Q_1 can create a power-dissipation problem for Q_1. Notice that when a series of zeners is used to get a given V_{REF} (Fig. 5–5b), the total zener voltage should be about 0.7 times the lowest input voltage.

A safe power-dissipation rating for the zener is three times the load. (With a 250-mV load, a zener capable of dissipating 750 mW should be good.) In the case of a regulator, the input resistance of the IC op-amp represents the load. Generally, IC op-amp resistance is quite high in relation to the voltage involved, so the power-dissipation rating of the zener is low. Typically, a zener (or series of zeners) used as a V_{REF} in an IC op-amp regulator requires a 0.5-W power-dissipation rating.

The value of R_S is found using the maximum input voltage, the desired zener voltage, and the safe power dissipation. Using the same 10- to 15-V input, a desired zener voltage (V_{REF}) of 7 V, and a safe power dissipation of 0.5 W, the first trial value of R_S is $(15 - 7)^2/0.5 = 128$ ohms. Use the next higher value of 130 ohms.

An alternative method of using a zener as V_{REF} is shown in Fig. 5–6. Here an FET is used in place of R_S. Note that the FET gate and source terminals are shorted together. With this arrangement, the FET conducts when the drop across the FET exceeds the pinch-off voltage. Typically, FET pinch-off is about 1 or 2 V (or less), so the input voltage need only be 1 to 2 V higher than the desired zener (or V_{REF}) voltage. This provides minimum drop across the series-pass transistor Q_1 (in Fig. 5–5).

5.5.2 Variable Voltage Reference Using an FET

One problem in using a zener as V_{REF} is that the reference voltage is fixed. The voltage depends on available zener values (either a single zener or a series of

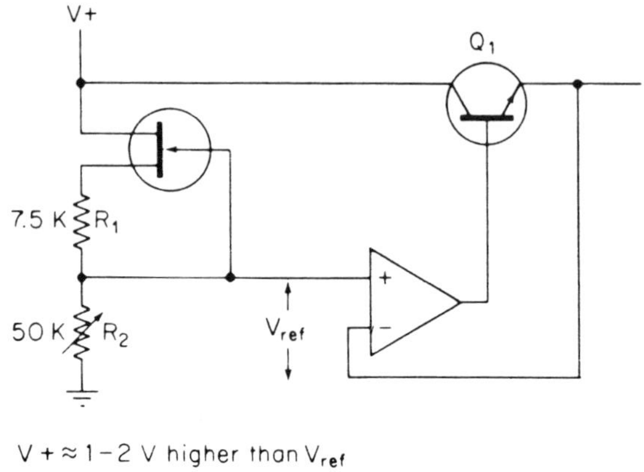

Figure 5–7. FET variable-voltage reference for op-amp regulator

zeners). Similarly, the reference voltage cannot be adjusted. Often it is better to have a regulator with a more flexible output. In a typical system, the power-supply output (the regulator output, in this case) should be capable of at least ±5% variation (preferably ±10%).

One inexpensive solution to this problem is shown in Fig. 5–7. Here the voltage reference is maintained at a selected value by series resistors and an FET. The values shown are typical for input voltages up to about 20 V and possibly higher. The input voltage must be higher than the highest desired V_{REF} by about 1 or 2 V (or less). This provides a minimum drop across the series-pass transistor Q_1.

Potentiometer R_2 is adjusted for the desired V_{REF}. Any variation from this voltage is applied to the FET gate and results in a corresponding variation in voltage drop across R_1. For example, if the input voltage goes up, V_{REF} goes up, as does the FET gate voltage. The FET draws more current, and the drop across R_1 increases to offset the initial change in V_{REF}.

Needless to say, the FET circuit of Fig. 5–7 cannot provide as stable a V_{REF} as the zener circuits of Figs. 5–5 and 5–6. However, when adjustable V_{REF} is of greater importance (and cost must be kept to a minimum), the circuit of Fig. 5–7 is satisfactory.

5.5.3 IC Regulators as a Voltage Reference

When V_{REF} must be adjustable and temperature stable, the packaged IC regulators (see Chapter 4) provide the best results. Figure 5–8 shows the basic circuit for using an IC regulator as V_{REF} in an op-amp regulator. As shown by the equation, V_{REF} is set by the ratio R_1/R_2 and a constant that is a characteristic of the IC regulator package.

Using a typical example, assume that the constant is 3.5 V, that R_2 is 10 kilohm, and that R_1 is variable from 10 to 40 kilohm. The V_{REF} is then variable from 7 to 17.5 V: $(1 + 10/10) \times 3.2 = 7$ V; $(1 + 40/10) \times 3.5 = 17.5$ V.

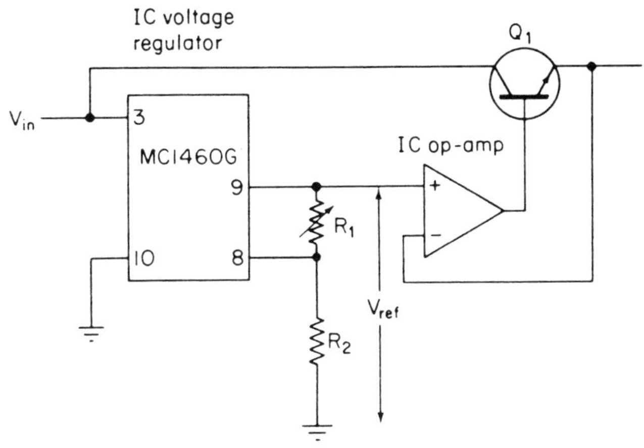

$$V_{ref} = \left(1 + \frac{R_1}{R_2}\right) 3.5\text{ V}$$

3.5 V = constant of IC voltage regulator

Figure 5–8. Using an IC regulator as V_{REF}

A typical temperature-drift specification for the IC regulator is 0.002%/°C. However, to realize the full value from the IC regulator, both R_1 and R_2 must have temperature coefficients that match those of the regulator.

5.6 Operating at Multiples of V_{REF}

Figure 5–9 shows the basic circuit used when the regulator output must be considerably higher than the available V_{REF} (operating at a multiple of V_{REF}). The regulator output voltage V_O is set by the R_2/R_1 ratio, plus 1, times V_{REF}. If R_2 is four times the value of R_1, V_O is five times V_{REF}, and so on.

Regulation for this circuit is not as good as that for a circuit where V_{REF} and V_O are equal, because of the loop attenuation introduced by R_1 and R_2. As shown by the equations, regulator output impedance Z_O, common-mode rejection CMRR, and power-supply sensitivity PSS are all degraded.

5.7 Floating IC Op-Amp Regulators

IC op-amp regulators are often used for stabilizing voltages considerably higher than the ratings of the IC. This control is possible because the entire circuit is not referenced to ground but "floated" between ground and the supply voltage. Such a circuit is shown in Fig. 5–10, and is essentially a *zener multiplier.*

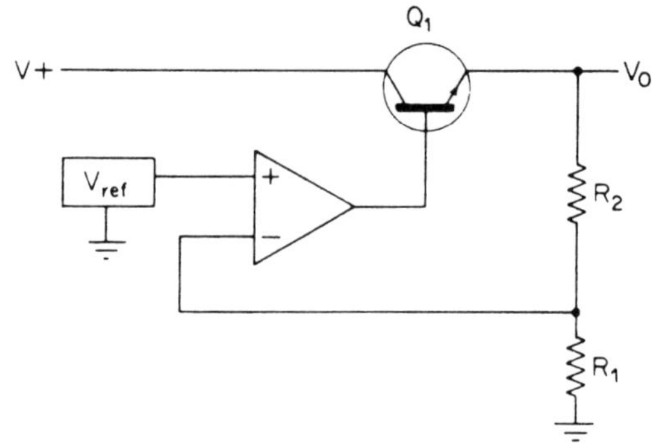

$$V_O \approx V_{REF}\left(1+\frac{R_2}{R_1}\right)$$

$$Z_O\text{ (closed loop)} \approx \frac{Z_O(\text{op-amp})}{\text{Beta } Q_1\left[1+\frac{A_{vol}\ R_1}{R_1+R_2}\right]}$$

$$\text{CMRR}' \approx \text{CMRR} \mid \text{dB} - \left(1+\frac{R_2}{R_1}\right) \mid \text{dB}$$

$$\text{Power supply sensitivity}' \approx \text{PSS}\left[1+\frac{R_2}{R_1}\right]$$

Figure 5-9. Operating at multiples of V_{REF}

The output V_O is a multiple of the reference voltage (across CR_3) and is set by the ratio R_1/R_2. Zener CR_1 ensures that the IC positive supply is greater than the required output swing. Zener CR_2 maintains a constant supply voltage for the op-amp. The voltage across CR_2 is the sum of the positive and negative supply voltages. Under these conditions, the voltage across the IC is tied to the output voltage by fixed constants. The voltage across CR_3 is equal to the output voltage less the zero common-mode level (which is the average of the V^+ and V^- voltage levels).

Typically, the voltage across CR_1 is 10 V, while the voltage across CR_2 is equal to the sum of the supply voltages (if V^+ and V^- are 15 V, CR_2 must be 30 V). Using these same values, and assuming a desired V_O of +75 V, the remaining value can be found using the equations of Fig. 5–10 as follows.

With a V_O of 75 V and a V_{CR_1} of 10 V, the value of V^+ is floated at +85 V, while the V^- floating point is +55 V. Note that the V^- point is 30 V (or the sum of the positive and negative supply voltages) below the V^+ point.

With the V^+ and V^- voltage levels (floating points) at +85 V and +55 V, respectively, the zero common-mode level is 70 V, or $(85 + 55)/2 = 70$ V. So both the inverting and noninverting inputs to the IC are floated at 70 V. With V_O at +75 V and the zero common-mode level of +70 V, the voltage across CR_3 is 5 V $(75 - 70 = 5)$.

Assume an arbitrary 1-mA current through R_1 and R_2, or a total of 75 kilohms for $R_1 + R_2$ (75 V/0.001 A = 75 kilohms). Using the equations, $R_1 = V_{CR_3}/0.001$ A,

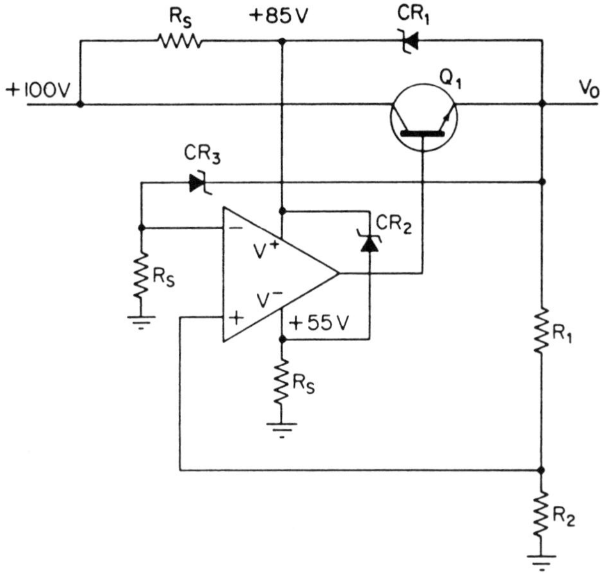

R_S = see text

$$V_O \approx \frac{R_1 + R_2}{R_1} \times VCR_3$$

$$R_1 \approx \frac{VCR_3}{0.001\,A}$$

$$R_2 = \left(\frac{V_O}{0.001}\right) - R_1$$

$$VCR_1 \approx 10V$$

$$VCR_2 \approx V^+ + V^-$$

$VCR_3 \approx V_O$ − zero common mode level

V^+ floating $\approx VCR_1 + V_O$

V^- floating $\approx V^+$ floating $- (V^+ + V^-)$

Zero common mode level $\approx \dfrac{(V^+ \text{ floating}) + (V^- \text{ floating})}{2}$

V^- V^+ = supply voltages of IC

V^- V^+ floating = level in reference to ground

Figure 5–10. Floating op-amp regulator

then R_1 = 5 kilohms (5/0.001 = 5 kilohms). With $R_1 + R_2$ at 75 kilohms and R_1 at 5 kilohms, R_2 = 70 kilohms.

The output V_O is $[(R_1 + R_2)/R_1] \times V_{CR_3}$, or $[(5 + 70)/5] \times 5 = +75$ V.

The value of R_1 is selected on the basis of voltage drop and power dissipation, as described for zener references in Section 5.5.1.

One limitation of the circuit in Fig. 5–10 is that regulation decreases (i.e., percentage of regulation increases) for increasing output voltage (all other factors being equal). Other limitations include a large voltage drop across R_2 and the R_S in series with CR_3, as well as across Q_1. However, the circuit can provide reasonably good regulation at voltages between about 100 and 250 V.

5.8 Effects of Ground Loops in Regulators

Thus far we have discussed the effects of devices (op-amp, zener, etc.) on overall regulator performance. In practical use, a well-designed regulator can perform very poorly if layout and *current paths* have not been considered carefully, particularly ground-current paths or ground loops.

Typically, the regulator circuit elements are mounted on the same board or card as the power supply. However, the regulator output can just as typically be delivered to a load at a remote location. (In this sense, the term *remote* can mean another board or card only a few inches away from the regulator.) In any event, the load current must pass through wires that produce some voltage drop.

Figure 5–11 shows the basic ground-loop model of an IC op-amp regulator. The critical load-current resistance paths are shown as RW_1 and RW_2 (remote wires 1 and 2). These resistances are the most important simply because the paths carry the most current, and thus drop the most voltage.

Even if the output voltage V_O equals V_{REF} and the IC op-amp gain is infinite (though neither condition is practical), the actual voltage across the load V_L is still a fraction of V_O. In effect, $V_L = V_O \times R_L/(RW_1 + RW_2 + R_L)$.

This may seem insignificant, but when you consider that No. 20 wire shows about 10 milliohm of resistance per foot, there is also about 1 mV per 100 mA of load current per foot. Assuming a 100-mA load (which is quite small for the typical power transistor used as a series-pass transistor Q_1), a V_{REF} of 10 V, and a distance of 6 inches from the regulator card to the load card (12 inches total), the actual V_O is offset by 1 mV from V_{REF}. This results in a best-possible 0.01% regulation.

An additional problem arises if the regulator output is connected to the load by binding posts or plug-in terminals rather than being soldered. Even solder joints can result in a loss of a few millivolts if not properly made.

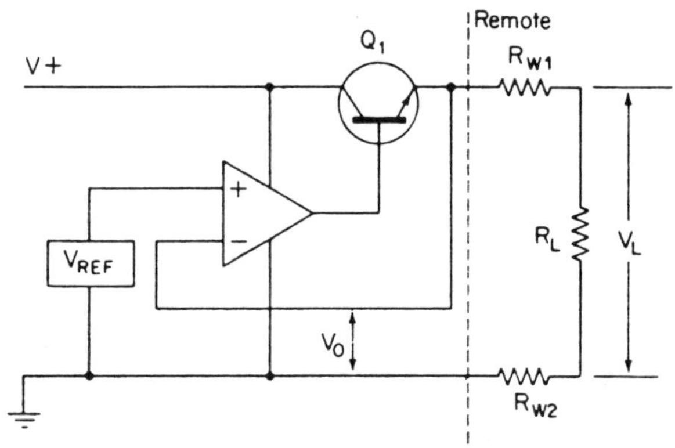

$$V_L \approx V_O \times \frac{R_L}{R_{W1} + R_{W2} + R_L}$$

Figure 5–11. Ground-loop model of an IC op-amp regulator

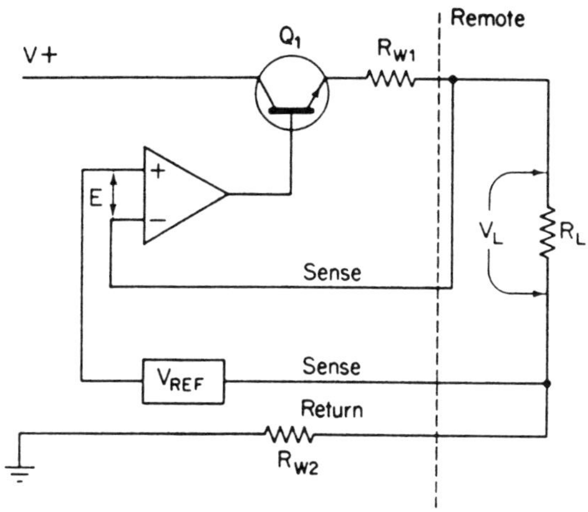

$$V_E^- = V_L + I_L R_{W2}$$

$$V_E^+ = V_{REF} + I_L R_{W2}$$

$$V_E^+ - V_E^- = E = V_{REF} - V_L, \text{ and}$$

$$\frac{V_E^+ + V_E^-}{2} = \frac{V_L + V_{REF}}{2} + I_L R_{W2}$$

Figure 5–12. Op-amp voltage regulator with remote sensing

5.8.1 Remote Sensing

The ground-loop problems can be minimized (but not eliminated) by remote sensing, as shown in Fig. 5–12. Assuming that the sense lines are the same length as the load lines, the voltages across the inverting and noninverting inputs to the IC are as shown by the equations in Fig. 5–12. In effect, both inputs are offset by the same amount.

In the circuit of Fig. 5–12, the common-mode voltage is increased by the drop across RW_2. If the common-mode rejection of the IC op-amp is good (at least 90 dB), the added common-mode voltage does not significantly affect performance of the regulator.

The sense lines should be No. 20 wire or smaller. (They need not be larger since they carry very little current.) Resistance RW_1 increases the open-loop output impedance of the regulator. However, the additional 10 to 20 milliohms (for a load-wire length of 1 foot) does not significantly affect performance.

5.9 Typical IC Op-Amp Voltage Regulator

Figure 5–13 shows a typical op-amp regulator circuit. This circuit incorporates both the IC regulator (see Chapter 4) and IC op-amp techniques discussed thus

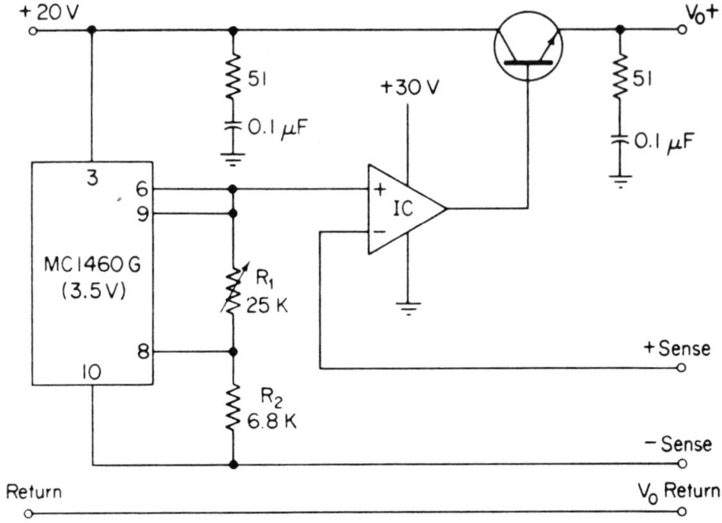

IC op-amp = MC 1539 G, or following characteristics
A_{Vol} = 50,000 (min)
Z_O = 4 K
CMRR = 100 dB
Offset voltage = −4 mV max
TC_{Vio} = 5 μV/°C
Power supply sensitivity = 150 μV/V (max)
Q_1 = 2N4921 with beta of 20 (min)

Figure 5–13. Typical op-amp and IC regulator

far. Figure 5–14 shows a typical regulation curve (the variation in output voltage at various load currents).

Note that the output voltage drops less than 0.25 V, with a load variation from 0 to 300 mA. Also note that the circuit components are identified as to type number or value. These values can be used as a starting point for simplified design. (Additional worked-out design examples are given in Chapter 7.)

The regulator operates from a +20-V source and is adjusted for a +15-V output by R_1. The drop across Q_1 is 5 V, and with a 300-mA load, Q_1 must be capable of dissipating at least 1.5 W (preferably more). The beta of Q_1 must be at least 20. The reference voltage supplied by the IC regulator to the IC op-amp is equal to the ratio of R_1/R_2 times a constant of 3.5.

The characteristics of the IC op-amp are listed in Fig. 5–13. Note that the IC op-amp is operated from a single supply (+30 V). For best results, R_1, R_2, and Q_1 should be mounted close together so that the components are all at the same temperature. This minimizes the change in percentage of regulation with changes in temperature. R_1 and R_2 should also be matched in temperature characteristics to the IC regulator and to each other.

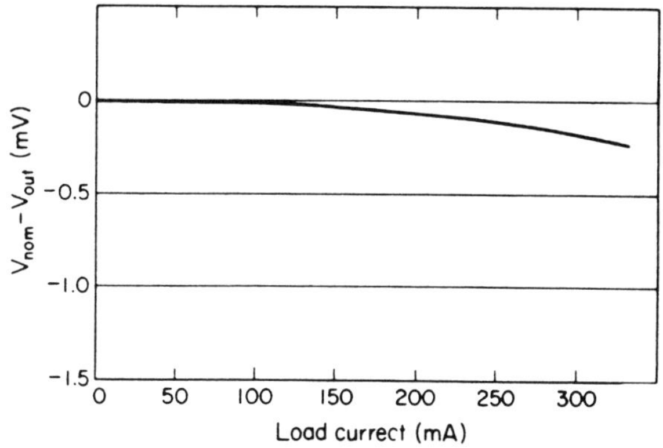

Figure 5-14. Load-regulation curve

In the final analysis, IC op-amp regulation (or linear regulation of any type) is no better than reference-voltage stability. In turn, the reference depends on temperature characteristics.

CHAPTER 6

Linear-Supply Testing and Connections

It is essential that you understand test procedures and connections for any phase of linear-supply work. The procedures can be applied to a just-completed linear supply during design/experimentation or to a suspected supply as part of troubleshooting.

The chapter is divided into three major sections. The first describes procedures that are generally sufficient for most practical applications (certainly for the home experimenter and serious hobbyist). The second describes some advanced test techniques that apply to a supply designed for commercial use. The chapter concludes with some notes on external connections for linear supplies, including remote sensing.

6.1 Linear-Supply Tests

The basic function of most linear supplies is to convert alternating current into regulated direct current. As discussed in Chapter 1, AC is first converted to DC by rectifiers (the basic power-supply function), then the DC is maintained at a desired voltage (or current) output (the basic regulator function). In any event, the linear-supply function can be checked quite simply by measuring the output voltage or current. However, for a thorough test of a linear supply, the output voltage (or current) should be measured without a load, with a load, and possibly with a partial load.

If the supply delivers the full-rated output voltage (or current) into a full-rated load, the basic function is met. In addition, it is often helpful to measure the regulating effect of a supply, the supply internal resistance, and the amplitude of any ripple at the linear-supply output. Also, because most linear-supply circuits depend on characteristics of a transformer, it is convenient (particularly in the experimental stage) to test the transformer as a separate component.

Sections 6.2 through 6.7 describe the basic linear-supply tests. Advanced tests are described in Sections 6.8 and 6.9. Linear-supply connections are covered in Section 6.10.

6.2 Basic Output Tests

Figure 6–1 shows the basic linear-supply test circuit. This arrangement permits the supply to be tested at no load, half load, and full load, depending on the position of load-selector switch S_1.

With S_1 in position 1, there is no load on the supply. At positions 2 and 3, there is half load and full load, respectively. The load resistors R_1 and R_2 must be noninductive so that a reactance is not placed on the supply output. Generally, the output load is considered as pure resistance for test purposes. Of course, if the supply must be operated with a reactive load, the supply can be so tested with a simulated reactance (capacitor or inductance) in addition to the pure resistance.

As shown by the equations of Fig. 6–1, the values of R_1 and R_2 are chosen on the basis of output voltage and load current (maximum or half load). For example, if the linear supply is designed for an output of 25 V at 500 mA full load, the value of R_2 is 25/0.5, or 50 ohms. The value of R_1 is 25/0.25, or 100 ohms.

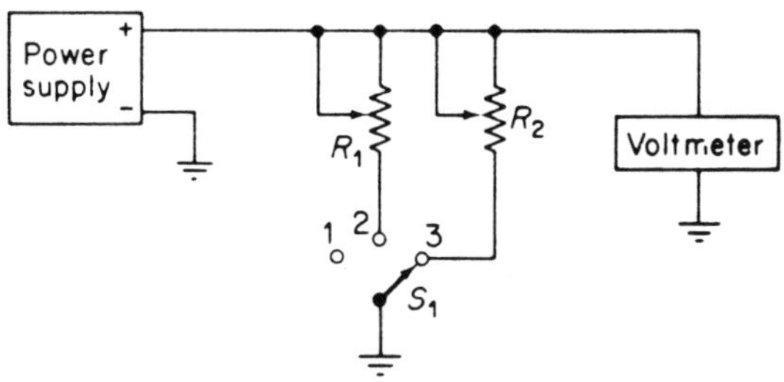

$$R_1 \text{ (in ohms)} = \text{half-load} = \frac{\text{rated output voltage}}{\frac{1}{2} \text{ max rated current}}$$

$$R_2 \text{ (in ohms)} = \text{full-load} = \frac{\text{rated output voltage}}{\text{max rated current}}$$

$$\text{Actual load current} = \frac{\text{voltage readout}}{\text{value of } R_1 \text{ or } R_2}$$

$$\%\ \text{regulation} = \frac{\text{no-load voltage} - \text{full-load voltage}}{\text{full-load voltage}} \times 100$$

$$\text{Power supply internal resistance} = \frac{\text{no-load voltage} - \text{full-load voltage}}{\text{current (amperes)}}$$

Figure 6–1. Basic linear-supply test circuit

Where more than one supply is to be tested, R_1 and R_2 should be variable and adjusted to the correct resistance value before test (using an ohmmeter with the power removed). If only one supply is to be tested, or as a temporary test setup, use fixed resistors. Either way, the resistors must be composition (not wire-wound) and must be capable of dissipating the rated power without overheating. For example, using the previous values for R_2 and R_1, power dissipation of R_2 is 25×0.5, or 12.5 W (use at least 15 W), and the dissipation for R_1 is 25×0.25, or 6.25 (use at least 10 W).

Use the following procedure to measure linear-supply output:

1. Connect the equipment as shown in Fig. 6–1.
2. If R_1 and R_2 are adjustable, set both to the correct values, using the equations.
3. Apply power. Set the input voltage to the correct value. Use the midrange value for the input voltage if a variable transformer or variac is available. For example, if the supply is designed to operate at 100 to 120 V, set the input to 115 V. (In most practical applications, simply use the available input supply voltage.)
4. Measure the output voltage at each position of S_1.
5. Calculate the current at positions 2 and 3 of S_1, using the equation for actual load current. For example, assume that R_1 is 100 ohms and the meter indicates 22 V at position 2 of S_1. The actual load current is 22/100, or 220 mA. If the supply output is 25 V at position 1 and drops to 22 V at position 2, this means that the supply does not produce full output with a load. This is an indication of poor regulation, possibly resulting from poor design (in the case of an experimental supply) or component failure (when discovered as part of troubleshooting).

6.3 Basic Output-Regulation Tests

Linear-supply output regulation is usually expressed as a percentage and determined by the following equation:

$$\%\ \text{regulation} = \frac{(\text{no-load voltage}) - (\text{full-load voltage})}{\text{full-load voltage}} \times 100$$

A low percentage-of-regulation figure is desired because this indicates that the output voltage changes very little with load. Use the following steps when measuring linear-supply output regulation.

1. Connect the equipment as shown in Fig. 6–1.
2. If R_2 is adjustable, set R_2 to the correct value, using the equation in Fig. 6–1.
3. Apply power. Measure the output voltage at position 1 (no load) and position 3 (full load).

4. Using the equation, calculate the percentage of regulation. For example, if the no-load voltage is 26 V and the full-load voltage is 25 V, the percentage of regulation is (26 – 25)/25 × 100, or 4% (very poor regulation).
5. Notice that the linear-supply regulation is usually poor (high percentage) when the internal resistance is high.

6.4 Basic Input-Regulation Tests

Linear-supply input regulation is usually expressed as a percentage and represents the maximum allowable output variation (with a given load) for maximum rated input variation. For example, if the supply is designed to operate with an AC input from 110 to 120 V and the DC output is 100 V, the output is measured (1) with an input of 120 V and (2) with an input of 110 V. If there is no change in output, input regulation is perfect (and probably impossible). If the output varies by 1 V, the output variation is 1% (which may or may not be within the allowable maximum).

The input regulation can be measured at full load and/or half load, as desired, using the same test connections shown in Fig. 6–1. However, the input voltage must be varied from maximum to minimum rated values (with a variable transformer or variac) and monitored with an accurate voltmeter.

6.5 Basic Internal-Resistance Tests

Linear-supply internal resistance is determined with the following equation:

$$\text{internal resistance} = \frac{\text{(no-load voltage)} - \text{(full-load voltage)}}{\text{current}} \times 100$$

A low internal resistance is most desirable because this indicates that the output voltage changes very little with load. Use the following procedure to measure linear-supply internal resistance.

1. Connect the equipment as shown in Fig. 6–1.
2. If R_2 is adjustable, set R_2 to the correct value, using the equation in Fig. 6–1.
3. Apply power. Measure the actual output voltage at position 1 (no load) and position 3 (full load).
4. Calculate the actual load current at position 3 (full load). For example, if R_2 is adjusted to 50 ohms and the output voltage at position 3 is 25 V (as in Section 6.3), the actual load current is 25/50, or 500 mA.
5. With the no-load voltage, full-load voltage, and actual current established, find the internal resistance. For example, with a no-load voltage of 26 V, a full-load voltage of 25 V, and a current of 500 mA (0.5 A), the internal resistance is (26 – 25)/0.5, or 2 ohms.

6.6 Basic Ripple Tests

Any linear supply, no matter how well regulated and filtered, has some ripple. This ripple can be measured with a meter or scope. Usually, the factor of most concern is the ratio between ripple and full output voltage. For example, if 3 V of ripple is measured, with a 100-V output, the ratio is 3 to 100, or 3% ripple (which is excessive).

6.6.1 Measuring Ripple with a Meter

Use the following procedure to measure output ripple with a meter.

1. Connect the equipment as shown in Fig. 6–1.
2. If R_2 is adjustable, set R_2 to the correct value, using the equation in Fig. 6–1.
3. Apply power. Measure the DC-output voltage at position 3 (full load).
4. Set the meter to measure alternating current. Any voltage measured under these conditions is ripple.
5. Find the percentage of ripple, as a ratio between the two voltages (AC/DC).

One problem often overlooked when measuring ripple is that the voltage is probably not a pure sine wave. Most meters provide accurate AC-voltage indications only for pure sine waves. A better method is to measure ripple with a scope, where the peak value can be measured directly.

6.6.2 Measuring Ripple with a Scope

A scope can be used to display the ripple waveform, from which the amplitude, frequency, and nature of the ripple voltage can be determined. Usually the scope is set to the AC mode when measuring ripple, since this blocks the DC output of the supply. Normally ripple voltage is small in relation to the supply voltage. If scope gain is set to display the ripple, the supply DC voltage drives the display off-screen.

1. Connect the equipment as shown in Fig. 6–2(a).
2. Apply power. If the test is to be made under load conditions, close switch S_1. Open S_1 for a no-load test. The value of R_1 must be chosen to provide the linear supply with the desired load.
3. Adjust the scope controls to produce two or three stationary cycles of each wave on the screen.
4. Measure the peak amplitude of the ripple on the scope's voltage-calibrated vertical scale.
5. Measure the frequency of the ripple on the scope's horizontal scale. When measuring ripple frequency, notice that a full wave produces two ripple "humps" per cycle, as shown in Fig. 6–2(b), whereas a half-wave supply produces one hump per cycle, as shown in Fig. 6–2(c).

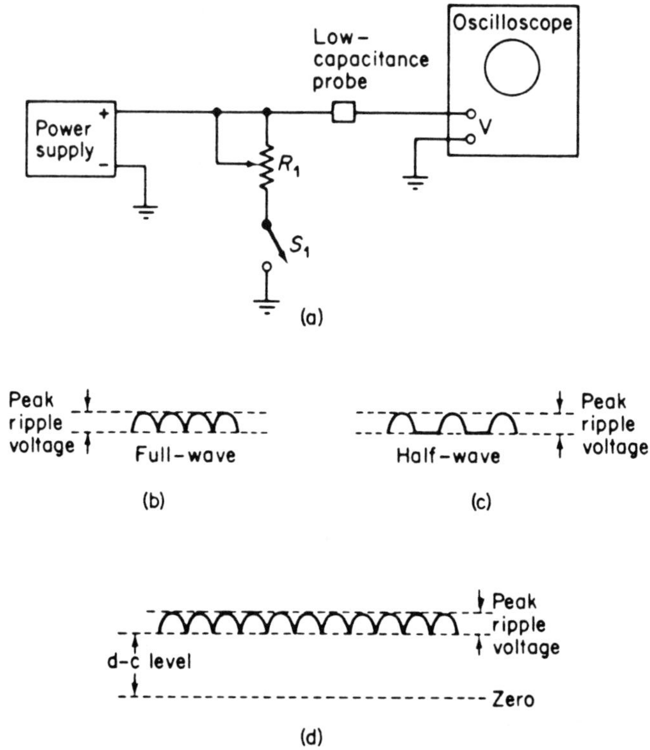

Figure 6–2. Measuring ripple with a scope (basic)

6. If the supply DC output must be measured simultaneously with the ripple, set the scope for DC mode. The baseline should be deflected upward to the DC level, and the ripple should be displayed at that level, as shown in Fig. 6–2(d). If this drives the display offscreen, reduce the vertical gain and measure the DC output. Then return the vertical gain to a level where the ripple can be measured and use the vertical-position control to bring the display back onto the screen for ripple measurement. This procedure works well for most lab scopes. On some shop-type scopes where the vertical-gain control must be set to a given "calibrate" position, the procedure is difficult, if not impossible. If this is the case, it is easier to measure the DC voltage with a meter.
7. Study the ripple waveform for defects in the linear supply (which can arise from poor design, poor layout, or component failure).

   - If the power supply is unbalanced (one rectifier passes more current than the others), the ripple humps will be unequal in amplitude.
   - If there is noise or fluctuations in the supply (particularly in zeners), the ripple humps will vary in amplitude or shape.
   - If the ripple varies in frequency, the AC source is varying.

- If a full-wave supply produces a half-wave output, one rectifier is not passing current.

6.7 Measuring Transformer Characteristics

If a linear-supply design does not prove satisfactory, the fault may be with the transformer. Although a supply component (such as a rectifier, capacitor, or IC regulator) is usually easy to replace during the experimental or design stage, a transformer can more easily be tested than replaced at this stage (unless another transformer is readily available for substitution).

The obvious test is to measure the transformer windings for opens, shorts, and the proper resistance value with an ohmmeter. In addition to basic resistance checks, it is possible to test a transformer's polarity markings, regulation, impedance ratio, and center-tap balance with a voltmeter.

6.7.1 Measuring Transformer Phase Relationships

When two supposedly identical transformers must be operated in parallel and the transformers are not marked as to phase or polarity, the phase relationship of the transformers can be checked using a voltmeter and power source. The test circuit is shown in Fig. 6–3. For power transformers, the source should be line voltage (110 to 120 V, or 220 to 240 V). Other transformers can be tested with a lower voltage dropped from a line or (in extreme cases) from an audio generator.

If the meter reading is zero or very low, the transformers are connected in the proper phase relationship. The transformers are out of phase if the secondary output voltage is double that of the normal secondary output. This condition can be corrected by reversing either the primary or secondary leads (but not both) of one transformer (but not both transformers).

If the meter indicates some secondary voltage, it is possible that one transformer has a greater output than the other. This condition results in considerable *local current* in the secondary winding and produces a power loss (if not actual damage to the transformers).

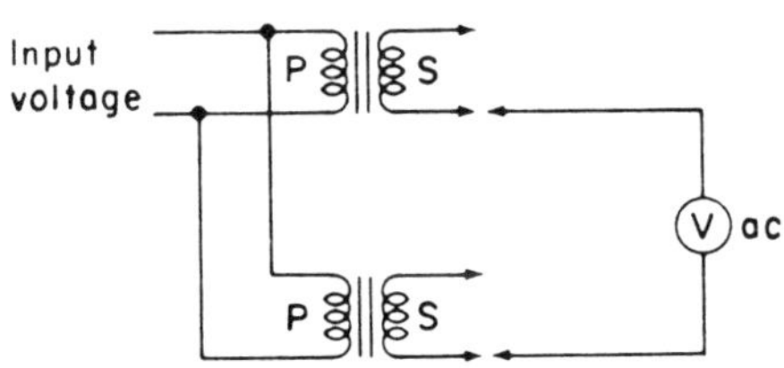

Figure 6–3. Measuring transformer phase relationships

6.7.2 Checking Transformer Polarity Markings

Many transformers are marked as to polarity or phase with dots, color-coded wires, or a similar system. Unfortunately, since transformer polarity markings are not always standard, they can prove very confusing during the experimental stage.

Generally, transformer polarities are indicated on schematics as dots next to the terminals. When standard markings are used, the dots mean that if electrons are flowing into the primary terminal with the dot, the electrons flow out of the secondary terminal with the dot. Thus, the dots have the same polarity as far as the external circuits are concerned. No matter what system is used, the dots or other markings show *relative phase,* because instantaneous polarities are changing across the transformer windings.

From a practical-test standpoint, there are only two problems of concern: the relationship of the primary to the secondary and the relationship of markings on one transformer to those on another.

The *phase relationship of primary to secondary* can be found using the test circuit of Fig. 6–4. First check the voltage across terminals 1 and 3, then across 1 and 4 (or 1 and 2). Assume that there is 3 V across the primary, with 7 V across the secondary. If the windings are as shown in Fig. 6–4(a), the 3 V is added to the 7 V and appears as 10 V across terminals 1 and 3. If the windings are as shown in Fig. 6–4(b), the voltages oppose each other and appear as 4 V (7 – 3) across terminals 1 and 3.

The *phase relationship of one transformer marking to another* can be found using the test circuit of Fig. 6–5. Assume that there is a 3-V output from the secondary of transformer A and a 7-V output from transformer B. If the markings are consistent on both transformers, the two voltages oppose, and 4 V is indicated. If the markings are not consistent, the two voltages add, resulting in a 10-V reading.

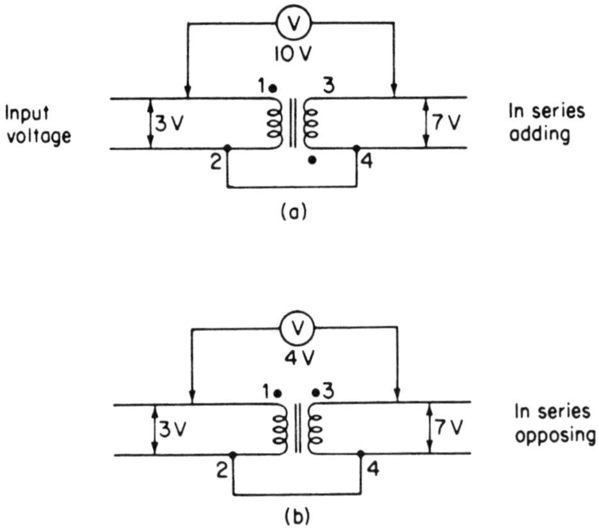

Figure 6–4. Checking transformer polarity markings

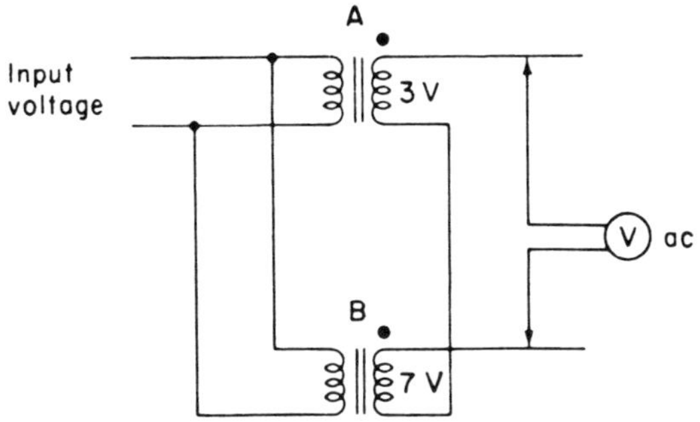

Figure 6–5. Checking transformer phase relationships

6.7.3 Checking Transformer Regulation

All transformers have some regulating effect, even though not used in a supply with a regulator circuit. That is, the output voltage of a transformer tends to remain constant with changes in load. Figure 6–6 shows the test circuit for finding transformer regulation and two equations for calculating regulation.

Whichever equation is used, remember that all transformers do not provide good regulation, even though designed for linear-supply use. The regulating effect produced by a transformer is determined by core material, winding arrangement, wire size, and so on. From a simplified-design standpoint, a transformer with very poor regulation requires extra filtering and/or regulation. Often, an inexpensive transformer shows poor regulation. Thus the savings on the transformer are offset by the increased cost of filter and regulator components.

In the test circuit of Fig. 6–6, the value of R_1 (load) should be selected to draw maximum rated current from the secondary. For example, if the transformer is to be used with a supply designed to deliver 100 W and the secondary output voltage is 25 V, the value of R_1 should be such that 4 A flows, or $R_1 = 25/4 = 6.25$ ohms.

First measure the secondary output voltage without a load and then with a load. Use either equation to find the percentage of regulation.

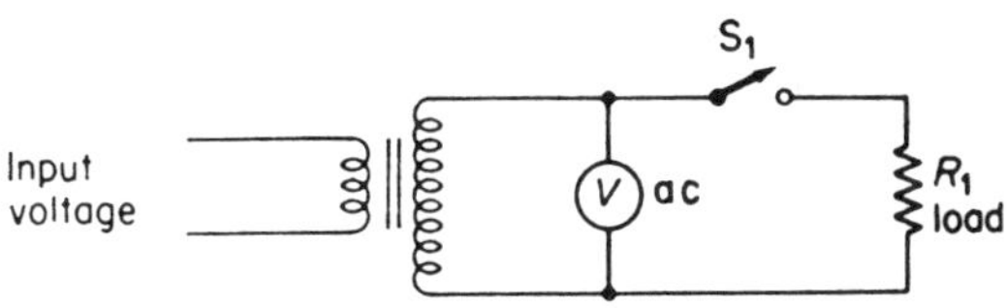

$$\%\ \text{regulation} = \frac{\text{no-load } V - \text{full-load } V}{\text{full-load } V} \times 100$$

or

$$\%\ \text{regulation} = \frac{\text{no-load } V - \text{full-load } V}{\text{no-load } V} \times 100$$

Figure 6–6. Checking transformer regulation

6.7.4 Checking Transformer Impedance Ratio

The impedance ratio of a transformer is the square of the winding ratio. For example, if the winding ratio is 15:1, the impedance ratio is 225:1. Any impedance value placed across one winding is reflected onto the other winding by a value equal to the impedance ratio.

Assuming an impedance ratio of 225:1 and an 1800-ohm impedance placed on the primary, the secondary then has a reflected impedance of 8 ohms. Similarly, if a 10-ohm impedance is placed on the secondary, the primary has a reflected impedance of 2250 ohms.

Impedance ratio is related to turns ratio (primary to secondary). However, turns-ratio information is not always available, so the ratio must be calculated using a test circuit as shown in Fig. 6–7. Measure both the primary and secondary voltage. Divide the larger voltage by the smaller, noting which is primary and which is secondary. For convenience, set either the primary or secondary to some exact voltage.

The turns ratio is equal to one voltage divided by the other. The impedance ratio is the square of the turns ratio. For example, assume that the primary shows 115 V, with 23 V at the secondary. This shows a 5:1 turns ratio (115/23) and a 25:1 impedance ratio (5^2).

6.7.5 Checking Transformer Winding Balance

There is always some imbalance in center-tapped transformers. That is, the turns ratio and impedance are not exactly the same on both sides of the center tap. A large imbalance can impair operation of any circuit, but especially a circuit such as that used in the power-supply section of a linear supply.

It is possible to find a large imbalance by measuring the DC resistance on both sides of the center tap. However, a small imbalance might not show up. It is usually

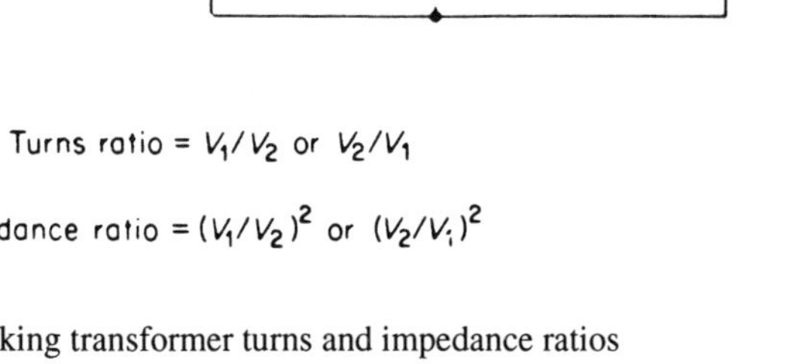

Turns ratio = V_1/V_2 or V_2/V_1

Impedance ratio = $(V_1/V_2)^2$ or $(V_2/V_1)^2$

Figure 6–7. Checking transformer turns and impedance ratios

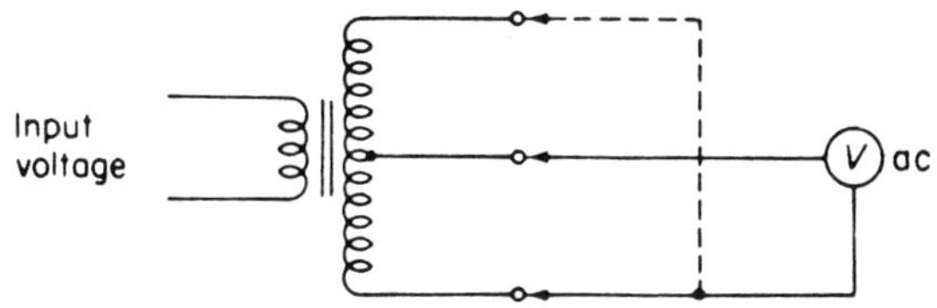

Figure 6–8. Checking transformer winding balance

more practical to measure the *voltage* on both sides of the center tap, as shown in Fig. 6–8. If the voltages are equal, the transformer winding is balanced. If a large imbalance is indicated by a large voltage difference, the winding can then be checked (with power removed, using an ohmmeter) for shorted turns, poor design, and so on.

6.8 Linear-Supply Advanced Tests

Figure 6–9 shows test connections suitable for measurement of the six most important operating specifications of a constant-voltage (CV) linear supply: source effect, load effect, periodic and random deviation (PARD), load-effect transient-recovery time, drift, and temperature coefficient. The automatic-load switch shown in Fig. 6–9 is used to interrupt the load periodically, when measuring transient-recovery time.

6.8.1 Test Precautions

The following precautions apply primarily to completely assembled and packaged linear supplies but can also pertain to supplies in the design/experimental stage.

Most of the tests described here can be performed with only four test instruments: variable autotransformer, differential or digital DC meter, true AC voltmeter, and oscilloscope.

Make certain that the autotransformer has an adequate current rating. If not, the input AC voltage applied to the supply may be severely distorted and the supply's rectifying/regulating circuits may operate improperly.

The DC voltmeter should have a resolution of 1 mV (or better), while the AC voltmeter should have a resolution of 1 mV full-scale. (Hewlett-Packard HP3420B, HP3455A, and HP3400A meet these requirements.)

The scope should have a sensitivity of 100 μV/cm and a bandwidth of 400 Hz for all measurements, except noise spike. A 5-mV sensitivity and a 20-MHz bandwidth are required for noise-spike measurements. (The HP180C with a 1821A time base and a 1806A vertical plug-in meets all requirements, except for noise spike. Use a 1803A plug-in with the HP180C for noise spike.)

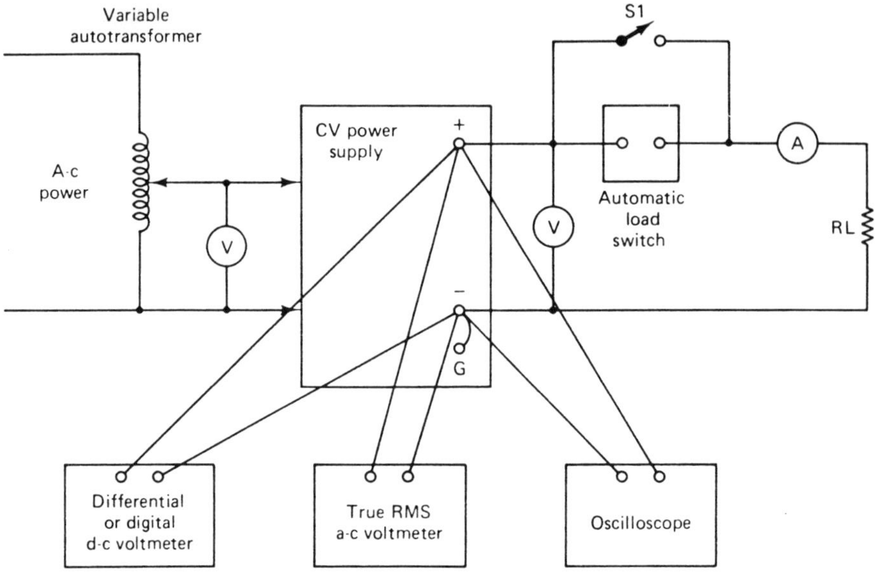

Figure 6–9. Advanced linear-supply test circuit

6.8.2 Test Connections

The connections shown in Fig. 6–9 apply to CV linear supplies where there are no remote-sensing leads. When remote-sensing (see Section 5.8.1) is involved, use the connections of Fig. 6–10. Either way, test connections should be permanent (not clip lead) and should be at the closest point to the supply (so that the monitoring device sees the same performance as the feedback circuit in the supply). Clip-lead connections often produce measurement errors because the contact resistance between the clip leads and supply terminals is typically greater than the supply output impedance.

All measurement instruments must be connected directly by *separate pairs* of leads. Twisted pairs or shielded cable should be used to avoid pickup on the leads. Here is a simple test for pickup and/or ground-loop effects. Turn off the supply and observe the scope for any unwanted signals (particularly at the line frequency) with the leads connected between the measurements (+S and –S, for example). Then connect both scope leads to whichever terminal is grounded to the supply. If there is any noise in either test condition, with the supply off, you have possible pickup and/or ground-loop effects.

6.8.3 Source Effect or Line Regulation

No matter what the test is called, the measurement is made by turning the variable autotransformer throughout the specified range from low-line to high-line and

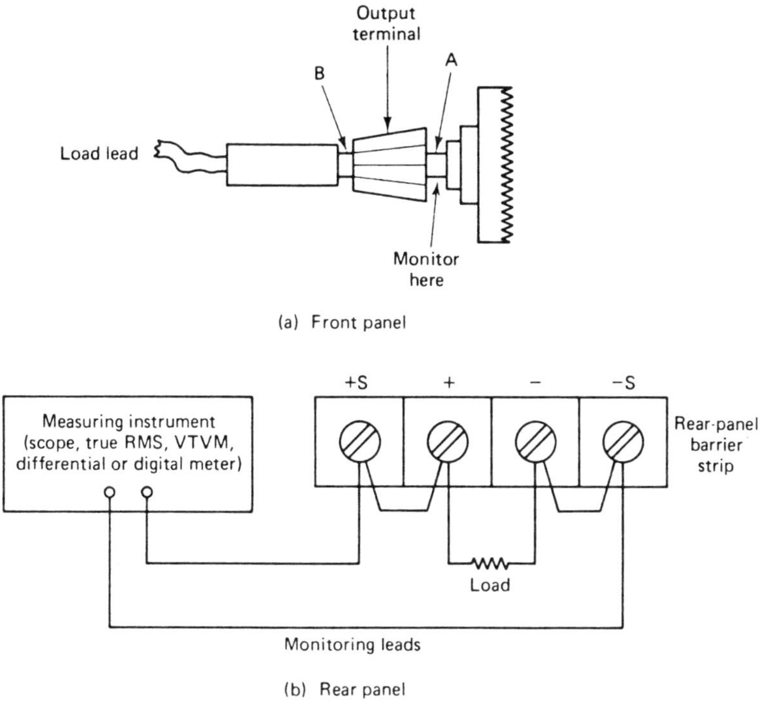

Figure 6–10. Test connections for remote sensing

noting the change in voltage at the output terminals. The test is performed with all other parameters constant. The supply should stay within specification for any rated output voltage, combined with any rated output current. The extreme source-effect test is with maximum output voltage and maximum output current.

6.8.4 Load Effect or Load Regulation

This measurement is made by closing and opening switch S_1 (Fig. 6–9) and noting the resulting static change in output voltage. The test is performed with all other parameters constant. The supply should stay within specifications for any rated output voltage, combined with any rated input line voltage. The extreme load-effect test is with maximum output voltage and maximum output current.

6.8.5 PARD or Noise and Ripple

In some commercial supplies, the term *PARD* (periodic and random deviation) has replaced the terms *ripple* and *noise.* When used, PARD represents deviation of the DC-output voltage from the average value, over a specified bandwidth, and with all other parameters constant.

Figure 6–11 shows some typical PARD (or noise and ripple) measurement circuits, together with some typical waveshapes. Notice that there are three possible waveshapes, together with the equations necessary to convert from one to the other. This is because the output ripple of a linear supply might resemble a square wave or a sawtooth wave more than a sine wave, even though the ripple is locked to the source or line frequency.

The connections shown in Fig. 6–11(a) produce a continuous *ground-loop* condition. Any current circulating in this loop causes a drop in series with the scope input. The resulting signal can easily be much greater than the true supply ripple and can completely invalidate the measurement.

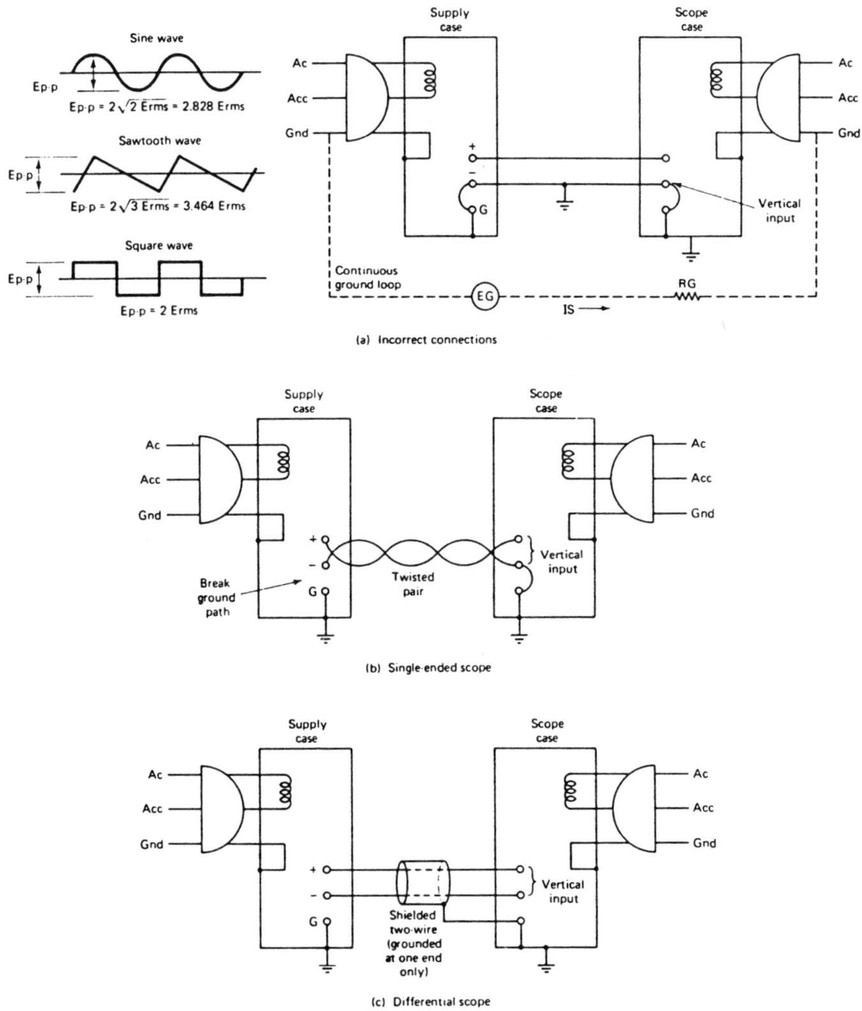

Figure 6–11. Typical PARD (or noise and ripple) measurement circuits

Figure 6–11(b) shows the correct method of measuring output ripple using a single-ended scope. The ground-loop path is broken by floating the supply output (breaking the ground path at the supply end). Use either a twisted pair or (preferably) a shielded two-wire cable. (Ground the shield at one end only to prevent ground currents from generating noise signals in the shielded leads.) To make sure that the scope is not displaying ripple induced in the leads, or picked up from grounds, short both scope leads together and touch both output terminals. If the ripple magnitude of the "shorted" test approaches the measurement ripple, the test is not reliable. Look for ground-loop or pickup problems.

In most cases the single-ended scope is satisfactory for all ripple and noise measurements. In certain cases, or for measurements where both the supply and scope case are connected to ground (for example, if both are rack mounted), the differential scope with a floating input is required. (Notice that two single-wire shielded cables may be substituted for the shielded two-wire cable.)

No matter what test connections are used, the equations in Fig. 6–11 can be helpful in converting the peak-to-peak scope display to the values given in the supply specifications. For example, if a supply is rated at a maximum of 1 mV (RMS), the maximum peak-to-peak ripple is 2.828 mV, assuming that the ripple is a sine wave. If the ripple is closer to a sawtooth, the maximum peak-to-peak is 3.464 mV. For a square-wave ripple, the maximum peak-to-peak is 2 mV. Of course, all of these values are approximations, and depend on the precise waveform.

6.8.6 Regulator Ripple Rejection and Output Resistance

Ripple rejection is a term applied to linear IC regulators (see Chapter 4) but not to the entire linear supply. Figure 6–12 shows typical test connections for measurement of ripple rejection. Although this circuit is designed specifically for a particular IC regulator (the Harris Semiconductor CA3085), variations of the circuit can be used with other regulators.

Notice that the circuit also provides a means of measuring output resistance for the regulator. The procedures for measuring both ripple rejection and output resistance are given in Fig. 6–12.

6.8.7 Regulator Noise Voltage

Figure 6–13 shows typical test connections for measurement of noise voltage. Again, these measurements apply to the IC regulator and not to the entire linear supply. Also, the circuit is designed specifically for the Harris Semiconductor CA3085 but can be used with other IC regulators (with changes in component values).

Notice that both the input (pin 3) and booster transistor (pin 2) of the IC are connected to a fixed 25 V. Any noise appearing at the output is being generated in the IC. However, if there is any noise from the 25-V source (ripple, noise spike, etc.), such noise can appear at the output.

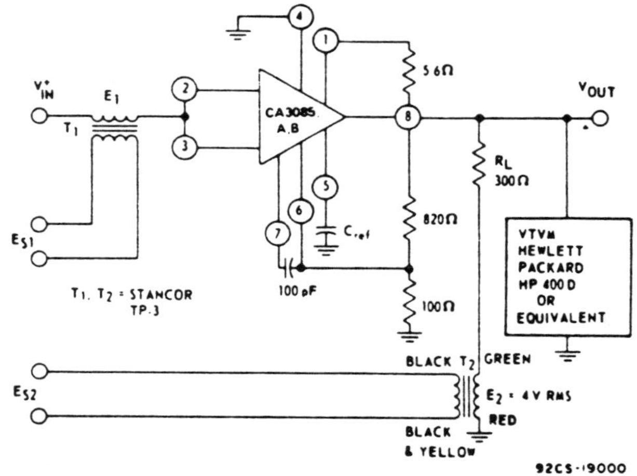

TEST PROCEDURES FOR TEST CIRCUIT FOR RIPPLE REJECTION AND OUTPUT RESISTANCE

Output Resistance

Conditions:

1. V_{IN} = +25 V, C_{REF} = 0, Short E_1
2. Set E_{S2} at 1 kHz so that E_2 = 4 V rms
3. Read V_{OUT} on a VTVM, such as a Hewlett-Packard, HP400D or equivalent
4. Calculate R_{OUT} from $R_{OUT} = V_{OUT}\,(R_L/E_2)$

Ripple Rejection — I

Conditions:

1. V_{IN} = +25 V, C_{REF} = 0, Short E_2
2. Set E_{S1} at 1 kHz so that E_1 = 3 V rms
3. Read V_{OUT} on a VTVM, such as a Hewlett-Packard, HP400D or equivalent
4. Calculate Ripple Rejection from 20 log (E_1/V_{OUT})

Ripple Rejection — II

Conditions:

1. Repeat Ripple Rejection I with C_{REF} = 2 μF

Figure 6–12. Measuring ripple rejection and output resistance (Harris Semiconductor, *Linear & Telecom ICs,* 1991, p. 2–43)

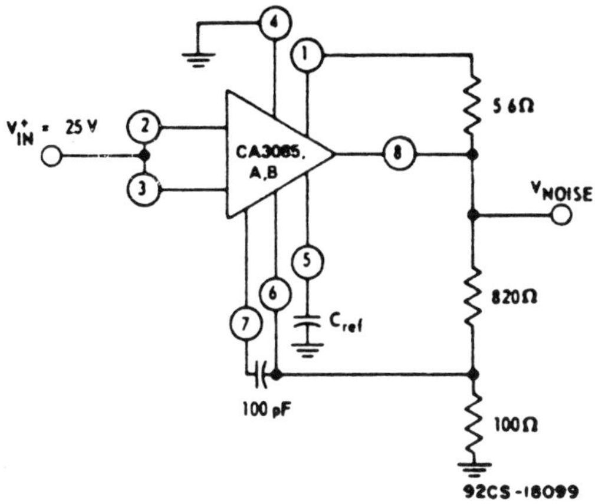

Figure 6–13. Measuring regulator noise voltage (Harris Semiconductor, *Linear & Telecom ICs,* 1991, p. 2–43)

6.8.8 Transient Recovery Time

Figure 6–14 shows the waveforms involved with transient recovery of a linear supply. If a step change in load current is imposed on the output, the output voltage shows a transient similar to that illustrated in Fig. 6–14. The output impedance of any supply rises at high frequencies, producing an *equivalent output inductance.* If

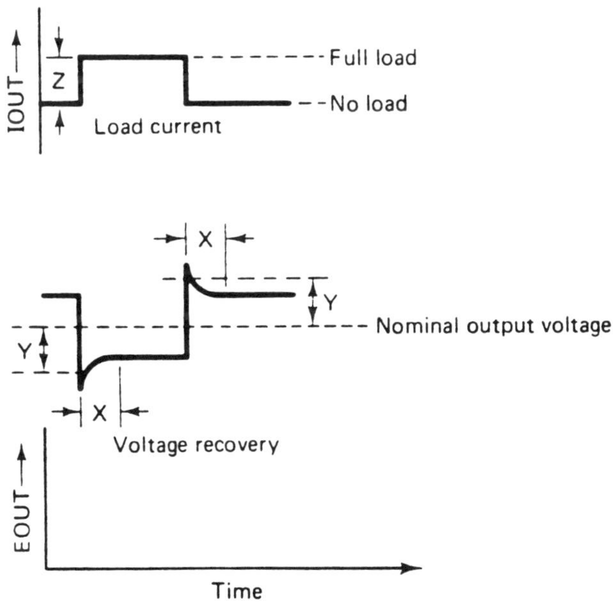

Figure 6–14. Waveforms involved with transient recovery of a linear supply

the load current is stepped or switched rapidly enough so that the high frequencies associated with the leading edge of the step change can react with this inductance, a spike occurs on the output terminals. So it is essential that rise time be established before specifying an amplitude for transient recovery.

As an example, a supply with an effective output inductance of 0.1 μH will show a load-transient spike of about 0.16 V if the load is switched with a rise time of 1 A/μs. However, the spike amplitude is only 160 μV if the load is switched at 1 A/ms. In the latter case, the output might not show up because the spike is small compared to the static change in output voltage associated with the full-load change.

Notice that Y shown in Fig. 6–14 is usually specified separately for each type of supply, but it is usually the same as the load-regulation specification (see Section 6.8.4). Z is the specified load-current change, but it is usually equal to the full-load current of the supply. The nominal output voltage is the DC level halfway between the steady-state output voltage before and after the imposed load change. Unless otherwise specified, transient recovery time may be measured at any input line voltage, combined with any output voltage and load current (within ratings).

Transient recovery time is not always defined in the same way. Sometimes the specification is omitted completely. In other cases, a given time is specified, but not the level or percentage of output-voltage recovery.

Figure 6–15 shows an automatic-load switch used by Hewlett-Packard to measure the load-effect transient recovery time of its supplies. The switch uses a mercury-wetted relay operated by a 60-Hz line voltage. (Use of the 60-Hz voltage makes it easy to observe on the scope.) The load-current duty cycle is controlled by R_1. The auto-load switch is connected into the test setup of Fig. 6–9. Do not exceed the maximum load ratings shown in Fig. 6–15 (to prevent damage to the relay contacts).

Figure 6–16 shows an alternate test circuit (including waveforms) for transient recovery time, developed by Harris Semiconductor. Instead of a mechanical relay, the load is applied and removed from the supply output by pulses applied to a transistor. Again, this circuit applies specifically to the IC regulator, but a similar circuit could be developed to test a complete supply.

6.8.9 Drift (Stability)

Drift is measured by monitoring the supply output on a differential or digital voltmeter over a stated measurement interval (typically, 8 hours after a 30-minute warmup). In some cases, a strip chart is used to provide a permanent record.

A thermometer should be placed near the supply to verify that the ambient temperature remains constant during the period of measurement. The supply should be at a location immune from stray air currents (away from open doors or windows and from air-conditioning vents). If possible, place the supply in an oven and hold at a constant temperature.

For a thorough test, measure both drift and PARD (or noise and ripple) over an 8-hour interval. (Make certain that the measuring instrument has a stability at least an order of magnitude better than the stability specification of the supply.)

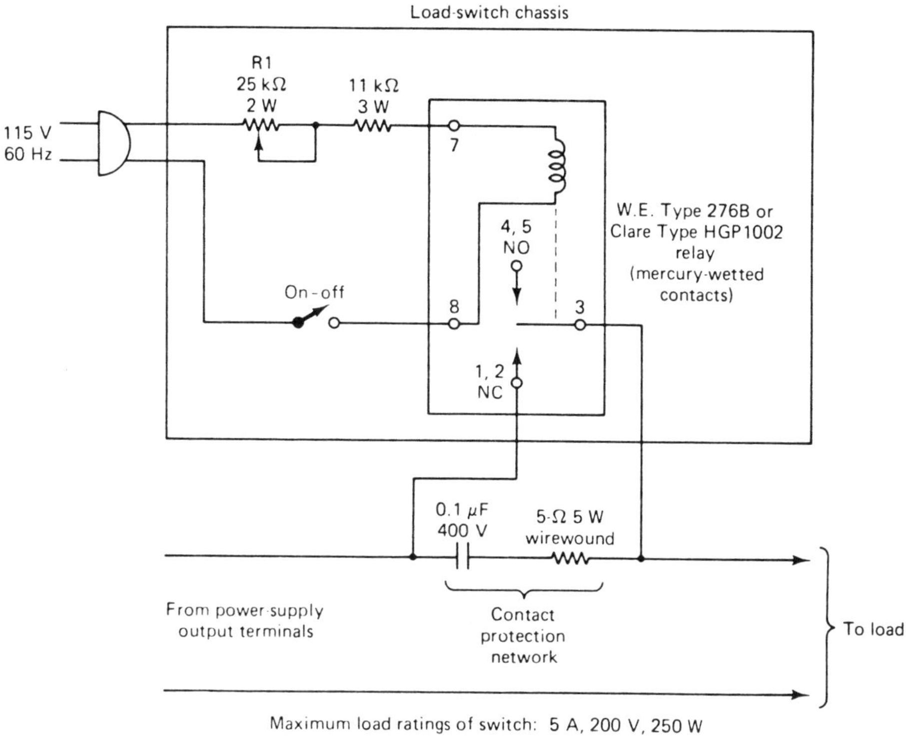

Figure 6–15. Automatic-load switch used to measure load-effect transient-recovery time

Typically, a linear supply drifts less over the 8-hour period than during the first 30-minute warmup period.

6.8.10 Temperature Coefficient

Temperature-coefficient measurements are made by placing the supply in an oven and varying the temperature over a given range, following a 30-minute warmup. The supply is allowed to stabilize at each measurement temperature. Although there is no standardization, a typical temperature coefficient specification is the change in output voltage per °C in ambient temperature, from 0°C to 55°C. The measuring instrument (digital or differential voltmeter) should be placed outside the oven and must have a long-term stability adequate to ensure that any voltmeter drift does not affect measurement accuracy.

6.9 Constant-Current-Supply Advanced Tests

Figure 6–17 shows test connections suitable for measuring the most important operating specifications of a constant-current (CC) linear supply. The automatic-

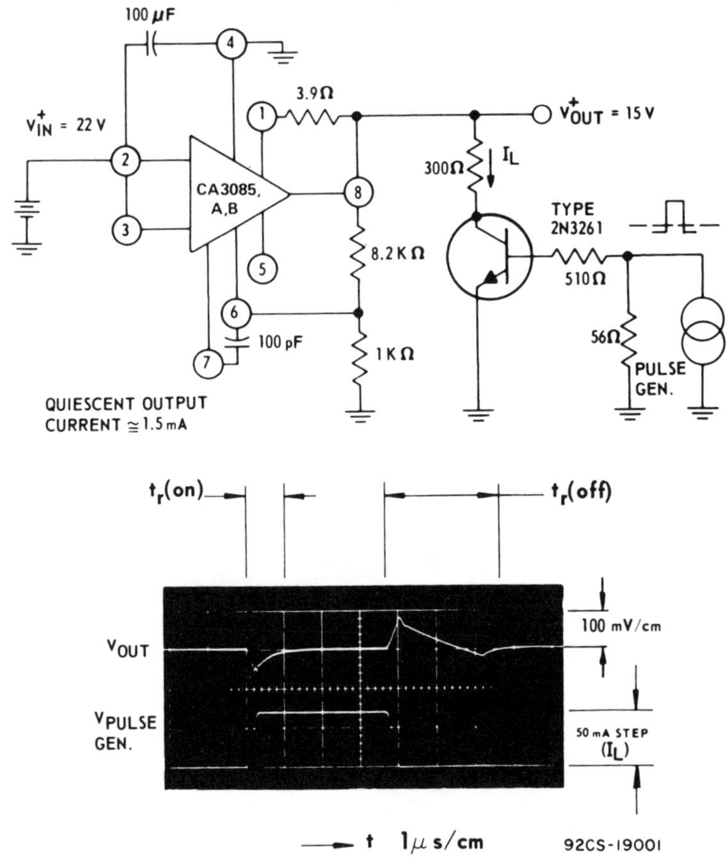

Figure 6–16. Alternate test circuit for transient-recovery time (Harris Semiconductor, *Linear & Telecom ICs,* 1991, p. 2–44)

load switch shown in Fig. 6–17 is used to interrupt the load periodically, when measuring transient-recovery time.

Notice that the procedures for CC supplies are similar to those of CV supplies, as discussed in Section 6.8. By comparing Fig. 6–17 to Fig. 6–9, one can see that there are three major differences between CC and CV measurements. First, the load switch is connected in parallel rather than in series with the supply load. This is because the supply performance is checked between short circuit and full load rather than between open circuit and full load.

Second, a current-monitoring resistor R_M is inserted between the supply output and the load. To simplify grounding problems, connect one end of R_M to the grounded output terminal of the supply.

Third, all CC measurements are made in terms of change in voltage across R_M. The CC performance is calculated by dividing these voltage changes by the resistance value of R_M.

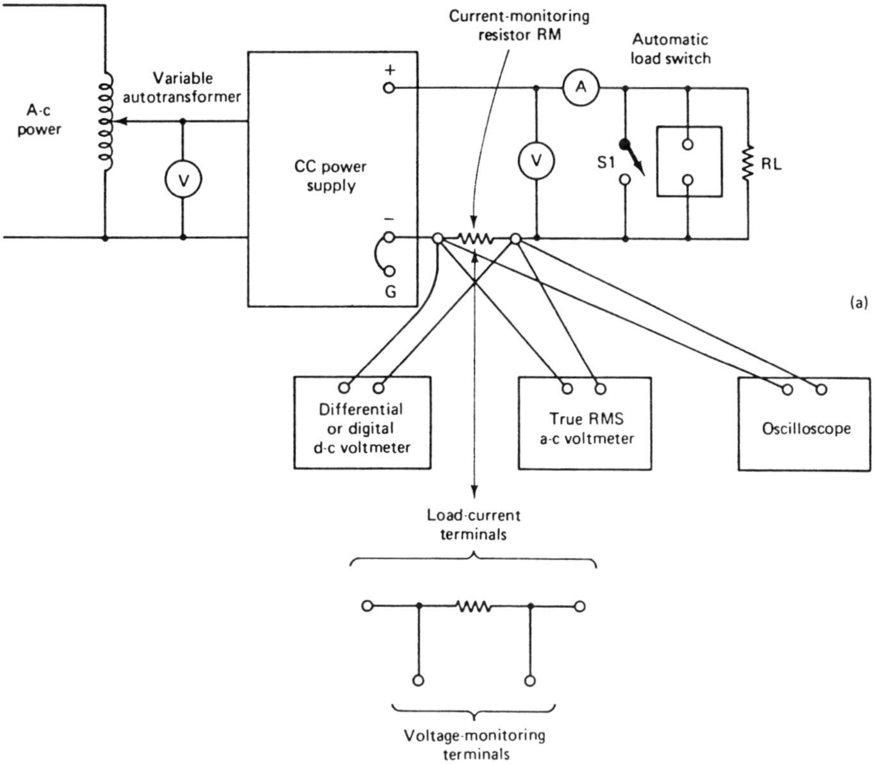

Figure 6–17. Constant-current linear-supply test circuit

6.10 Typical Linear-Supply Connections

Although most linear-supply connections are relatively simple, certain precautions must be observed. Such factors as ground loops, load decoupling, common connections, load-wire sizes, current limits, oscillation, and remote sensing must be considered.

If these connection factors are ignored, even a well-designed supply can produce excessive output ripple, a tendency toward oscillation, poor load and line regulation, and unnecessary degradation of stability, temperature coefficient, and transient-recovery time. If carried to the extreme, incorrect linear-supply connections can result in total failure and/or shock hazard! Careful attention to the guidelines in this section (developed by Hewlett-Packard for its equipment) will improve the safety and usefulness of any linear supply.

6.10.1 Power Cords

Typically, a power cord has three wires: AC, acc (a common or ground), and third-wire safety ground. These wires must never be interchanged.

If the acc and ground wires are interchanged, the supply chassis is placed at the acc potential. This causes ground currents to flow through the chassis and ground-return paths and can cause excessive ripple as well as other malfunctions.

If the AC and acc wires are interchanged, any supply switches or fuses are placed in series with the ground side of the power line, instead of the hot side. If the line switch is turned off, or the fuse opens, the hot side of the line is connected to the exposed components within the supply.

If the AC and safety-ground wires are interchanged, the chassis may be raised to the level of the line voltage. This can result in potentially lethal shock hazard if the chassis is not grounded. If the chassis is grounded, the fuses will probably blow!

If you cannot tell the difference between the AC and acc lines at the power outlet, use a continuity checker or voltmeter and measure between each line and ground. The hot lead is AC, while the cold lead is acc. (Of course, if you cannot tell the difference between AC and acc, you probably should not be working with linear power supplies!)

6.10.2 Isolation Transformers

If an isolation transformer (or an autotransformer) is used between the power outlet and supply, the transformer should be rated at 200% of the maximum RMS current required by the supply. If not, the transformer may become saturated and prevent the supply from delivering the full output voltage and current. If an autotransformer is used, make certain that the transformer common terminal is connected to acc (and not to AC). Otherwise, the supply acc terminal has an abnormally high voltage (which can cause ripple and/or shock hazard).

6.10.3 Line Regulators

Line regulators are generally not recommended for linear supplies. The problem is even greater for switching supplies and/or supplies with SCR preregulation. Line regulators often change the shape of the output waveform, thus offsetting any improvement produced by a constant line input to the supply. If line regulators are used during the experimental stage, the test results may be erroneous.

6.10.4 Line-Wire Sizes

Make certain that the wire size for the line is rated to carry at least the maximum linear-supply input current. If there is a long run between the outlet and supply, it may also be necessary to consider the impedance of the line cord or wires. For example, some supplies fail to function properly if the IZ drop in the line cord approaches 10% of the line voltage, even though the wire size is sufficient for maximum current and the voltage at the supply is within the rated load or value. As a general guideline, the line cord should have wire sizes so that the IZ drop is below 1% of the nominal line voltage (at maximum rated supply input current).

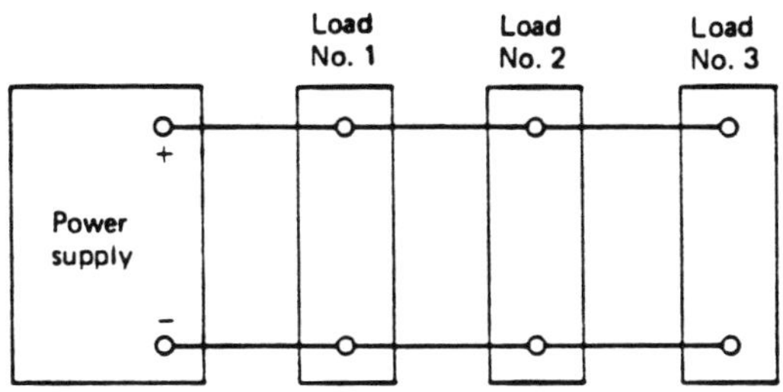

Figure 6–18. Common mistake when connecting a single linear supply to multiple loads

6.10.5 Load-Line Connections

Figure 6–18 shows the most common mistake when connecting a single linear supply to multiple loads. Under these conditions, each load sees a supply voltage that depends on the current drawn by other loads (and, to a lesser extent, by the IZ drop caused by the load leads). If any of the loads draws a variable current, there can be interaction between the loads (possibly resulting in unstable voltage across the loads, oscillation, noise, etc.).

Figures 6–19 and 6–20 show the preferred method for connecting a single supply to multiple loads. With this system, a single pair of terminals is used for all connections from the supply. Notice that the DT terminals (Fig. 6–20) can be at the supply output, at the load, or at a separate pair of terminals established expressly for distribution. The choice depends on the location of the remote-sensing terminals (if any). Remote sensing is discussed in Section 5.8.1, and internal connections are shown in Figs. 6–12 and 6–13.

If there is no remote sensing (Fig. 6–19), locate the DTs as close as possible to the supply output terminals (or use the supply output terminals, if practical, for best performance).

If remote sensing is used (Fig. 6–20), locate the DTs as close as possible to the load terminals. Then connect the supply sensing terminals to the DTs.

With either remote or nonremote sensing, connect one pair of wires directly from the supply output terminals to the DTs and a separate pair of wires from the DTs to each load. There should be no direct connection from one load to another, except by way of the DTs, even if the supply output terminals are used as the DTs.

Notice that Figs. 6–19 and 6–20 also show connections from a single supply to a single load when the supply has sensing terminals. If there are no sensing terminals, the supply output terminals should still be connected directly to the load. For best results, the load lines should be twisted to prevent (or minimize) pickup from stray magnetic fields. Also, the sensing lines should be shielded.

6.10.6 Load-Wire Ratings

There are two basic factors that determine the size or rating of the load wires. First, and most obvious, each load wire must be of sufficient size to carry the supply

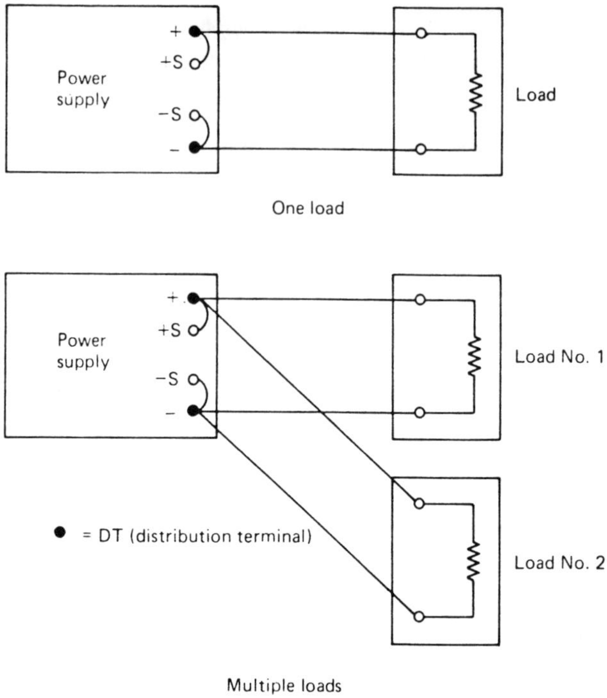

Figure 6–19. Connecting a single supply to multiple loads (without remote sensing)

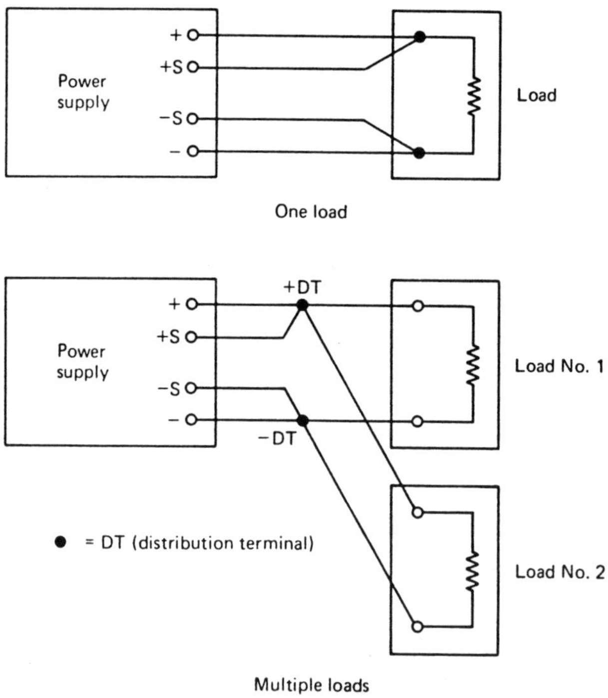

Figure 6–20. Connecting a single supply to multiple loads (with remote sensing)

Table 6–1. Resistance and current ratings for various wire sizes

AWG (B&S) Wire Size	*Annealed Copper Resistance at 20°C (mΩ/ft)*	*Nominal Current Rating (A) Single Conductor in Free Air at 30°C with Rubber or Thermoplastic Insulation*
22	16.1	5
20	10.2	7
18	6.39	10
16	4.02	13
14	2.53	20
12	1.59	25
10	0.999	40
8	0.628	55
6	0.395	80
4	0.249	105
2	0.156	140
0	0.0993	195
00	0.0779	260

output current that flows if the *load terminals are short-circuited.* Second, not so obvious, is that the supply and load wires both have impedance. Fortunately, the impedances are not of great practical importance, except when the load wires are very long. It is only when the load-wire impedance is equal to (or greater than) the supply output impedance that problems arise (oscillation in amplifier circuits, false triggering in digital and pulse circuits, etc.).

Table 6–1 shows resistance and current ratings for various wire sizes. These can be used to calculate the best wire sizes for given linear-supply loads, as well as possible current flow under short-circuit conditions.

6.10.7 Load-Decoupling Capacitors

One way to minimize the effects of high-frequency impedance is to use a decoupling capacitor at the load. This is shown in Fig. 6–21, which illustrates the preferred method for connecting more than one supply to multiple loads. The capacitors reduce the high-frequency impedance seen by any individual load (look back toward the supply) and reduce mutual-coupling effects between loads fed from the same supply.

Load-decoupling capacitors are most often used with multiple loads that draw pulse currents with short rise times (square-wave pulses, etc.). Without decoupling,

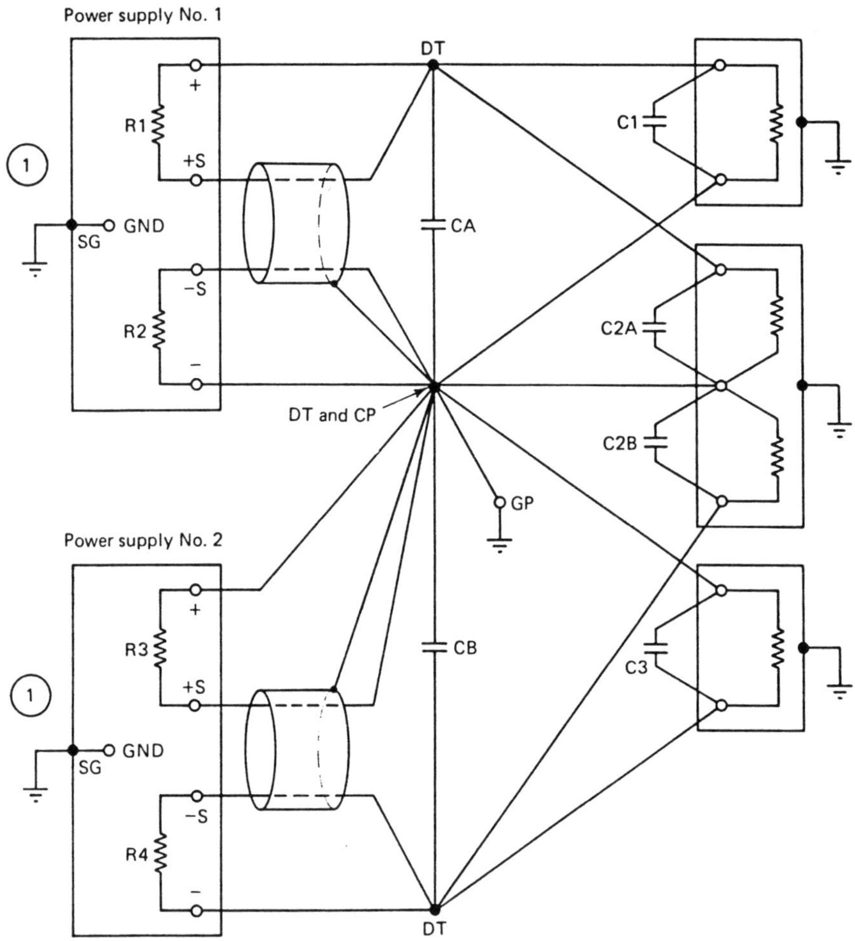

(1) = Power-supply chassis ground connection via third-wire safety-ground lead and rack frame
SG = Safety-ground lead in power cord
GND = Power-supply ground terminal
CA, CB = Power-supply output capacitors removed and placed across DTs
R1, R2, R3, R4 = Remote-sensing protection resistors

Figure 6–21. Connecting more than one supply to multiple loads

these current changes can cause spikes that travel down the load-distribution lines and falsely trigger one of the other loads.

To be effective, the high-frequency impedance of the decoupling capacitors must be lower than the impedance of the wires connected to the same load. Also, use the shortest possible leads to connect the capacitors directly to the load and DTs so that the wire impedance between the capacitor and connecting point is minimized.

6.10.8 Ground-Loop Problems

With few exceptions, the problems caused by ground loops are of much greater importance than load-wire impedance. No two ground points (or GPs, Fig. 6–21) have exactly the same potential (even two points on the same PC board). Any potential difference (even a fraction of a volt) causes current to flow through a complete ground loop (which can be defined as any circuit with more than one ground point).

No matter what the definition, to avoid ground-loop problems there must be *only one ground-return point* in a linear-supply system (the linear supply and all loads, and all other linear supplies connected to the same loads). This is not always easy, particularly in large systems that tend to force compromises with the ideal grounding concept.

Consider the following example of a rack-mounted system. Such systems usually consist of separately mounted supplies and loads (with multiple ground connections). Each instrument usually has its own chassis tied to the third-wire safety ground load of the power cord. The rack is often connected by a separate wire to the safety ground (a cold-water pipe).

Under these conditions, with instrument panels fastened to the rack frames, circulating ground currents are inevitable. However, as long as the ground currents are *confined to the ground system* and do not flow through any portion of the supply DC-distribution wiring, the effect on system performance is probably negligible.

If there is only one connection between the ground system and distribution system, ground-loop currents do not affect the supply DC-output and load circuits. However, any magnetic coupling or capacitive coupling between the two systems can provide a return path to link both systems. Such coupling can be minimized by means of common-point (CP) and ground-point (GP) concepts, as shown in Fig. 6–21.

The resistors R_1 through R_4 in Fig. 6–21 provide protection for the sensing circuits. The resistors ensure that there is some voltage between the sensing terminals, even if there is an opening in the sensing lines (because of broken wires, switches, relays, etc.) If a sensing opening occurs, the regulator circuit reacts as though the load voltage is zero. Usually the output voltage corrects this deficiency by climbing rapidly toward the maximum rectifier voltage (a value significantly higher than the supply maximum-rated output voltage). Even if the supply is designed to withstand this extreme, the load may not be so fortunate! The resistors prevent such an open-circuit condition, even if the sensing wires are disconnected.

CHAPTER 7

Linear-Supply Design Examples

All the design examples for linear power supplies in this chapter use off-the-shelf components, and all the design techniques discussed thus far are used as needed. These circuits can be used immediately as is or, by altering component values, as a basis for simplified design of similar linear supplies. Since the present trend is to use some form of IC regulator in linear supplies, we concentrate on such circuits here, starting with a simple voltage reference.

7.1 IC Voltage Reference

Figures 7–1, 7–2, and 7–3 show the connection data, basic circuit wiring, and performance characteristics for a typical voltage reference (the Raytheon REF–02). As shown in Fig. 7–2, the REF–02 is essentially a three-terminal regulator (see Chapter 5), where the output can be adjusted over a 5-V ± 300-mV range.

The REF–02 contains a bandgap reference using thin-film resistors, a step-up amplifier, short-circuit protection, and a zener-trim network. The output remains at 5 V ± 0.3% in the presence of large changes of temperature, load current, and input voltage.

An adjustment or trim pin is provided to change the output by at least 3% with little effect on temperature coefficient. A tempco pin provides a voltage that varies linearly with temperature, typically from +470 mV to +830 mV over the military temperature range. The adjustable output allows the system designer to trim system errors by setting the reference to a voltage other than 5 V. Of course, the output can also be set to exactly 5.000 V or to 5.12 V for binary operation. Adjustment of the output does not significantly affect the temperature performance of the IC. Typically the temperature-coefficient change is 0.7 ppm/°C for 100 mV of output adjustment.

Because the REF–02 is a simple device, there is no need for elaborate design examples. However, one point must be considered when operating with input voltages greater than about 22 V. As shown in Fig. 7–3(a), the maximum load current is

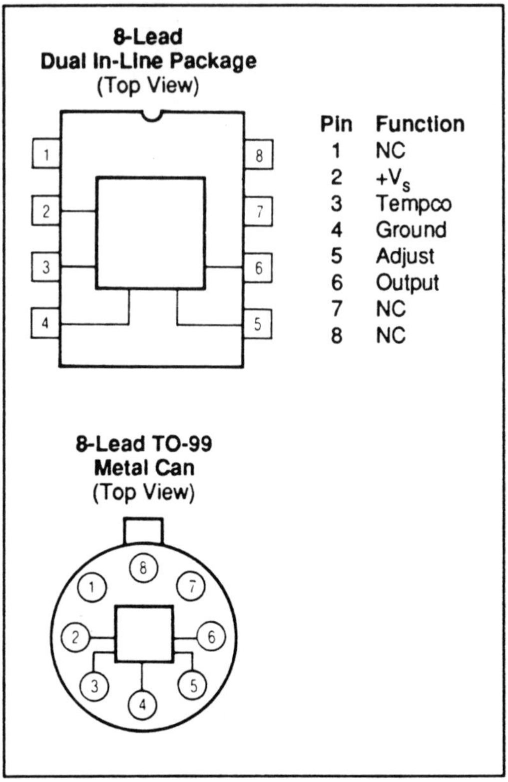

Figure 7–1. REF–02 voltage reference connection data (*Raytheon Linear Integrated Circuits,* 1989, p. 8–9)

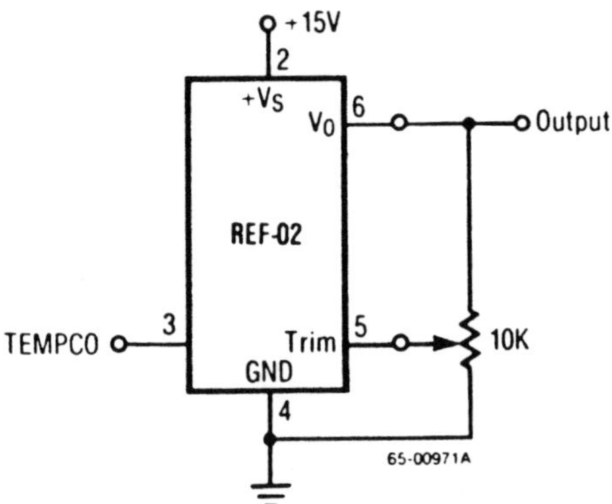

Figure 7–2. REF–02 voltage reference basic circuit wiring (*Raytheon Linear Integrated Circuits,* 1989, p. 8–14)

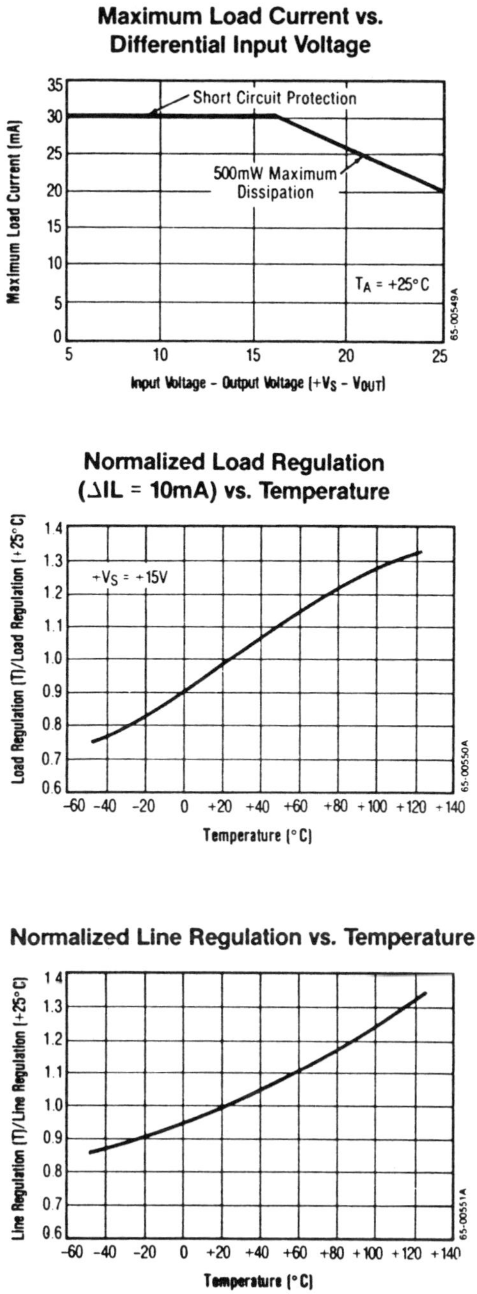

Figure 7–3. REF–02 voltage reference performance characteristics (*Raytheon Linear Integrated Circuits,* 1989, p. 8–14)

30 mA. However, this begins to drop to 20 mA when the differential between input and output voltages exceeds about 17 V. (The rated input voltage is from +7 V to +40 V.)

7.2 Adjustable Three-Terminal Regulator

Figure 7–4 shows the functional schematic of a three-terminal regulator (the classic National Semiconductor LM117) capable of producing an output voltage from 1.2 V to 27 V at 1.5 A. As shown, the circuit is similar to the op-amp regulators discussed in Chapter 5. The op-amp and biasing circuits are arranged so that all quiescent current is delivered to the regulator (rather than to ground), eliminating the need for a separate ground terminal. A 1.2-V reference voltage appears inserted between the noninverting input of the op-amp and the adjustment terminal. About 50 μA is needed to bias the reference. This current comes from the adjustment terminals.

In operation, regulator output is the adjustment-terminal voltage plus 1.2 V. If the adjustment terminal is grounded, the device acts as a 1.2-V regulator. For higher output voltages, a divider, R_1 and R_2, is connected from the output to ground as shown in Fig. 7–5. The 1.2-V reference across R_1 forces 5 mA of current to flow. This 5 mA then flows through R_2, increasing the voltage at the adjustment terminal (and therefore the output voltage). The output voltage is given by

$$V_{OUT} = 1.2\ V\left(1 + \frac{R_2}{R_1}\right) + 50\ \mu A \cdot R_2$$

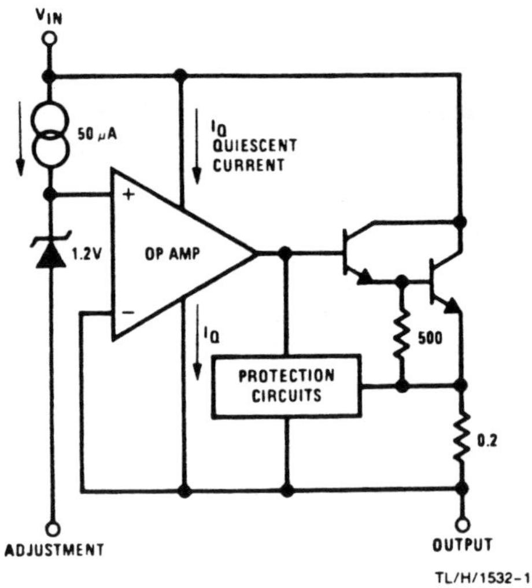

Figure 7–4. Functional schematic of LM117 three-terminal regulation (National Semiconductor, *Linear Applications Handbook,* 1991, p. 448)

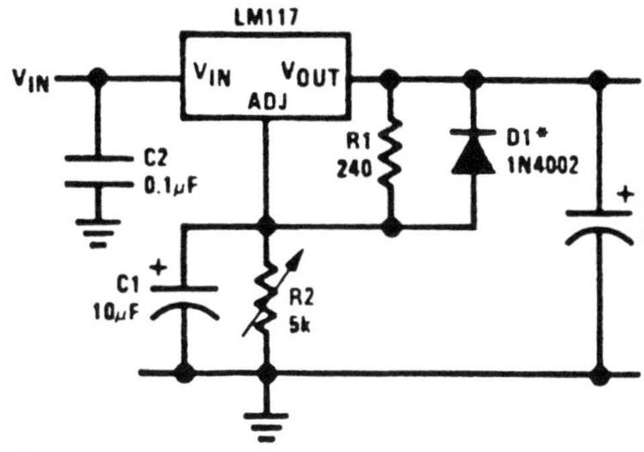

Figure 7–5. Adjustable three-terminal regulator (National Semiconductor, *Linear Applications Handbook,* 1991, p. 448)

The 50-μA biasing current is small compared to 5 mA and causes only a small error in actual output voltages. Of course, programming currents other than 5 mA can be used, depending upon the application. Since the regulator is floating, all quiescent current is absorbed by the load. With too light a load, regulation is impaired. Usually the 5-mA programming current is sufficient. However, worst-case minimum load for commercial-grade parts requires a minimum load of 10 mA.

7.2.1 Overload-Protection Circuits

The LM117 has both current-limit and thermal-overload protection. Current limit is set internally at about 2.2 A and remains constant with temperature. Thermal-overload protection, included on the chip, turns the regulator off when the chip temperature exceeds about 170°C. Also included is safe-area protection for the pass transistor to decrease the current limit as input-to-output voltage differential increases. The safe-area protection allows full output current at 15 V differential and does not allow the current limit to drop to zero at high input-to-output differential voltages. This prevents start-up problems with high input voltages.

7.2.2 Basic Regulator Operation

The basic regulator shown in Fig. 7–5 requires two resistors and a standard input capacitor. Resistor R_2 sets the output voltage while R_1 provides the 5-mA programming current. The two capacitors on the adjustment and output terminals are optional for improved performance. The 10-μF bypass capacitor provides 80-dB ripple rejection (at 120 Hz), while the output capacitor prevents excessive ringing. If these values are used and the output voltage does not exceed 25 V, no protective diodes are required. Figure 7–6 shows an LM117 with protective diodes included for use with outputs greater than 25 V and high values of output capacitance.

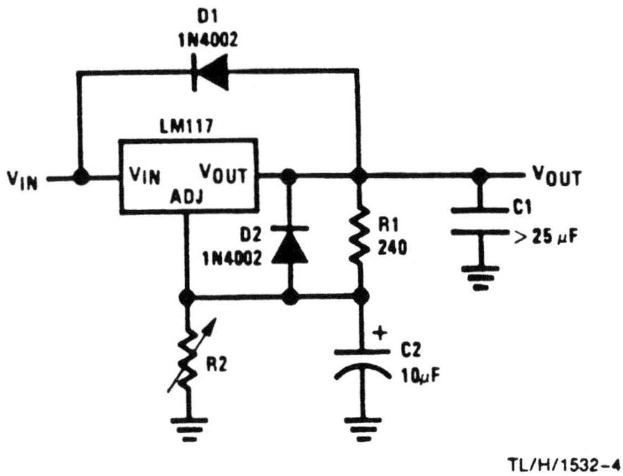

$$V_{OUT} = 1.25V\left(1 + \frac{R2}{R1}\right) + R2 \bullet I_{ADJ}$$

D1 protects against C1 (input shorts)

D2 protects against C2 (output shorts)

Figure 7–6. Regulator with protective diodes (National Semiconductor, *Linear Applications Handbook,* 1991, p. 449)

Some care should be taken in making connection to the LM117 to achieve the best load regulation. Series resistance between the regulator output and programming resistor R_1 should be minimized. Any voltage drop because of load current through this series resistance appears as a change in reference voltage and degrades regulation. If possible, two wires should be connected to the output: one wire for the load current and one for R_1. The ground of R_2 can be returned near the ground of the load to provide remote sensing and improved load regulation.

7.2.3 Five-Volt Logic Regulator with Electronic Shutdown

Figure 7–7 shows how an LM117 can be shut down by a TTL signal. When the TTL signal is applied, the transistor saturates, R_2 is shorted out, and the output drops to about 1.2 V. Under these conditions, most 5-V circuits will draw only a small fraction of the normal operating current and, in effect, will be turned off.

7.2.4 Adjusting Multiple On-Card Regulators with Single Control

Figure 7–8 shows a method of adjusting multiple on-card regulators so that all outputs track within ±100 mV. The adjustment terminals of all devices are tied together and a single divider is used to set the outputs. Programming current is set at 10 mA to minimize the effects of the 50-μA bias current of the regulators and

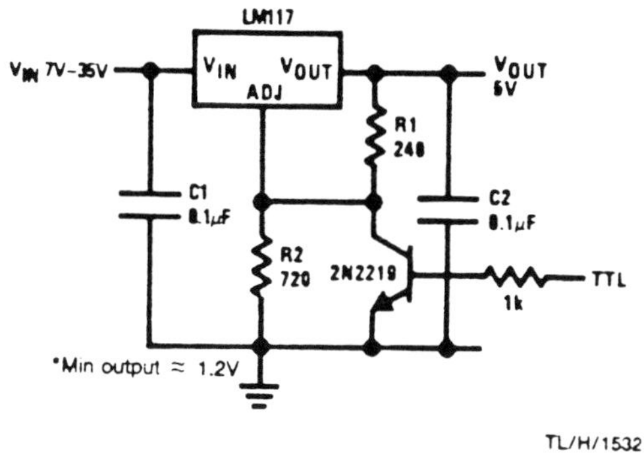

Figure 7–7. Five-volt logic regulator with electronic shutdown (National Semiconductor, *Linear Applications Handbook,* 1991, p. 450)

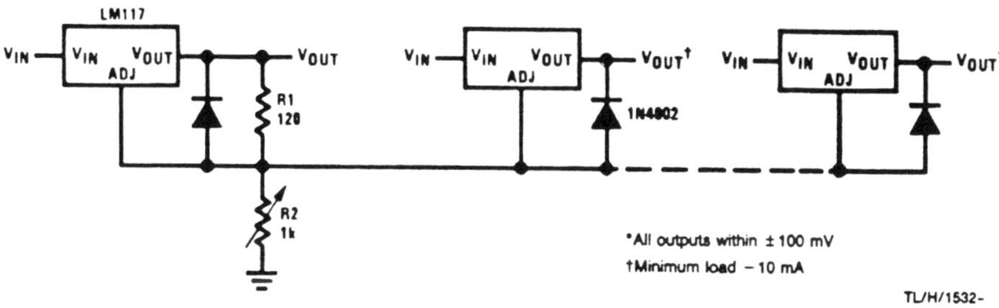

Figure 7–8. Adjusting multiple on-card regulators with single control (National Semiconductor, *Linear Applications Handbook,* 1991, p. 450)

should be further increased if many LM117s are used. The diodes connected across each regulator ensure that all outputs will decrease if one regulator is shorted.

7.2.5 Combined Regulator and Voltage Reference

Figure 7–9 shows an LM117 used as a 15-V regulator while the programming current powers an LM127 zener. The LM127 is an IC zener with less than 1 ohm dynamic impedance, and it can operate over a range of 0.5 mA to 15 mA with virtually no change in performance.

7.2.6 Regulator with Multiple Outputs

Figure 7–10 shows how the programming current of an LM117 can be used to provide multiple reference voltages. Here the output-setting resistor (R_1 through R_4) is tapped to provide multiple output voltages to op-amp buffers. An additional transis-

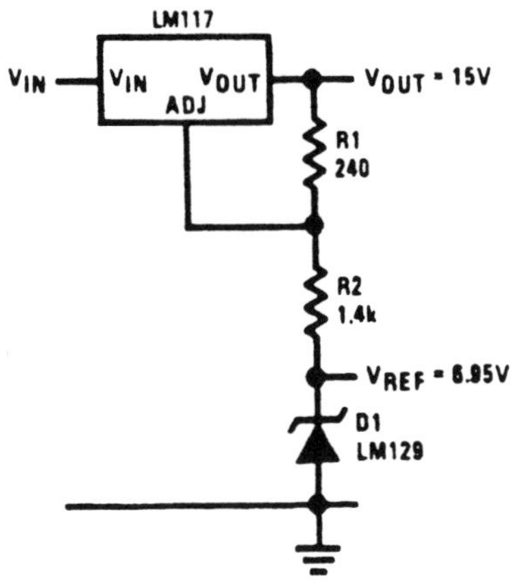

Figure 7-9. Combined regulator and voltage reference (National Semiconductor, *Linear Applications Handbook,* 1991, p. 441)

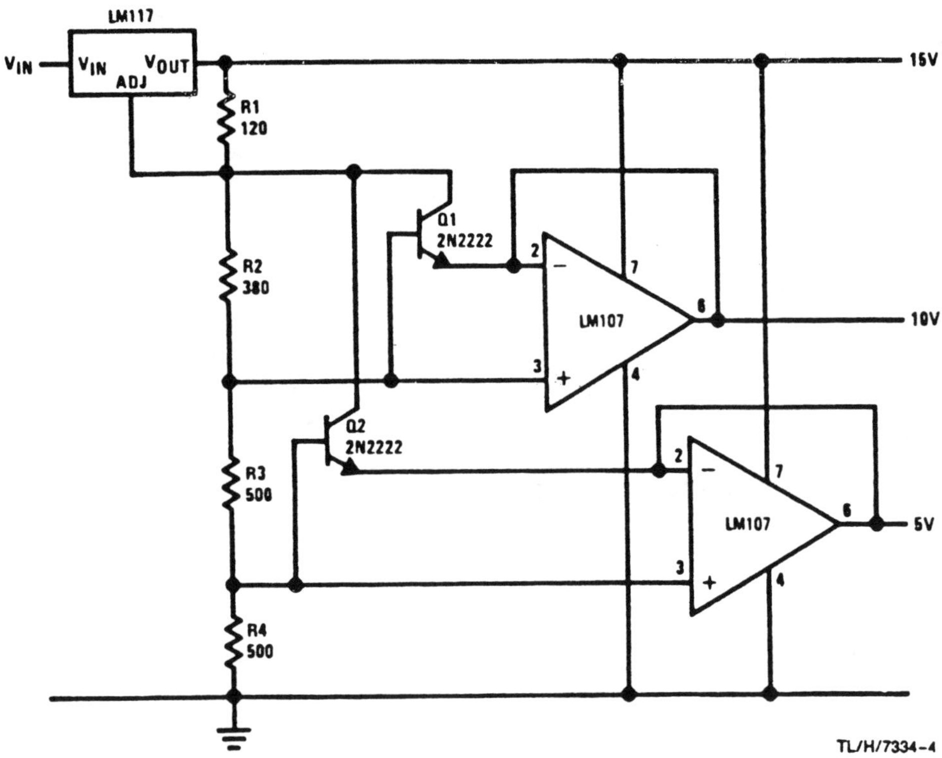

Figure 7-10. Regulator with multiple outputs (National Semiconductor, *Linear Applications Handbook,* 1991, p. 442)

tor is included as part of the overload protection. When any output is shorted, the op-amp will current-limit, and a voltage will be developed across the input. This will turn on the transistor and pull down the adjustment terminal of the LM117, causing all outputs to decrease, minimizing possible damage to the rest of the circuit.

7.2.7 Two-Terminal Current Regulator

Figure 7–11 shows how a three-terminal voltage regulator, such as the LM117, can be converted to a two-terminal current regulator. The output current is set by the regulator's internal reference and the value of R_1. Use a value slightly higher than the internal reference when making the calculation. (In this example use 1.25 V, which is slightly higher than the 1.2 V of the LM117.) Any load current can be selected within the minimum and maximum limits discussed in Section 7.2 (10-mA worst-case minimum and 1.5-A maximum). For example, to get a constant 0.8-A output current, use 1.5625 (or the nearest standard) for R_1.

7.2.8 Adjustable CV/CC Regulator

Figure 7–12 shows how an adjustable-current regulator is combined with a voltage regulator to form a CV/CC lab-type supply. The current is sensed by R_1, while R_2 sets the current-limit point. When R_2 is adjusted, a portion of the LM117 1.2-V reference is canceled by the drop across R_2, decreasing or increasing the current-limit point. At low output currents, current regulation is degraded since the voltage across R_1 becomes quite low. For example, with 50 mA output current, only 50 mV is dropped across R_1 and the supply rejection of the LM117 will limit the current regulation to about 3%, for a 40-V change across the LM117. Figure 7–13 shows an alternate current regulator using an additional LM117 to provide the reference (rather than D_1, as shown in Fig. 7–12). Note that the current regulators of Figs. 7–12 and 7–13 require a negative supply to operate down to ground.

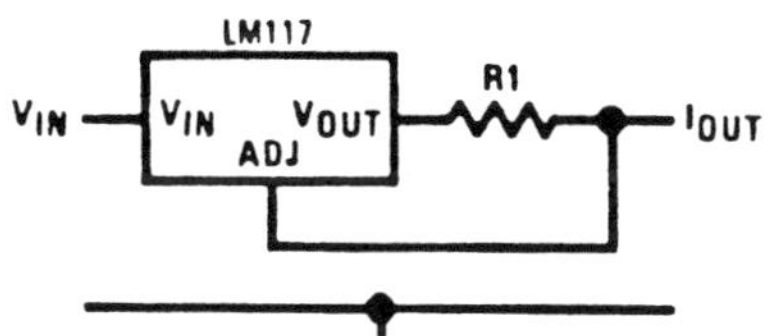

$$I_{OUT} = \frac{1.25V}{R1}$$

$$10\ mA \le I_{OUT} \le 1.5A$$

Figure 7–11. Two-terminal current regulator (National Semiconductor, *Linear Applications Handbook,* 1991, p. 442)

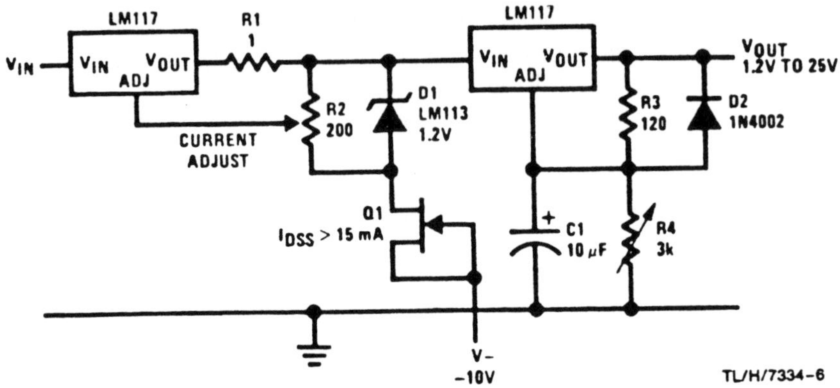

Figure 7–12. Adjustable CV/CC regulator (National Semiconductor, *Linear Applications Handbook,* 1991, p. 442)

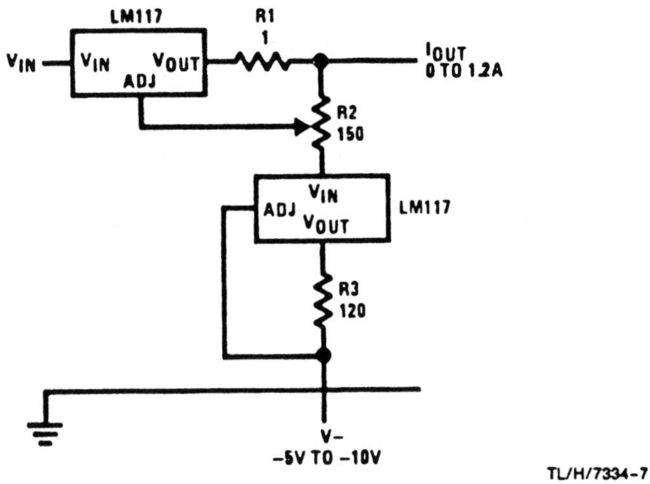

Figure 7–13. Alternate regulator using an additional three-terminal regulator as the reference (National Semiconductor, *Linear Applications Handbook,* 1991, p. 443)

7.3 Basic Fixed Three-Terminal Regulator

Figure 7–14 shows the characteristics and basic connections for a typical fixed three-terminal regulator (the classic National Semiconductor LM109). Because of the virtually foolproof operation and single purpose, fixed three-terminal regulators do not require much applications information. However, the following paragraphs describe some points that are often overlooked when designing with fixed regulators.

Typical Characteristics of the Logic-Card Regulator: $T_A = 25°C$

Parameter	Conditions	Typ
Output Voltage		5.0V
Output Current		1.5A
Output Resistance		0.03Ω
Line Regulation	$7.0V \le V_{IN} \le 35V$	0.005%/V
Temperature Drift	$-55°C \le T_A \le 125°C$	0.02%/°C
Minimum Input Voltage	$I_{OUT} = 1A$	6.5V
Output Noise Voltage	10 Hz ≤ f ≤ 100 kHz	40 μV
Thermal Resistance Junction to Case	LM109H (TO-5) LM109K (TO-3)	15°C/W 3°C/W

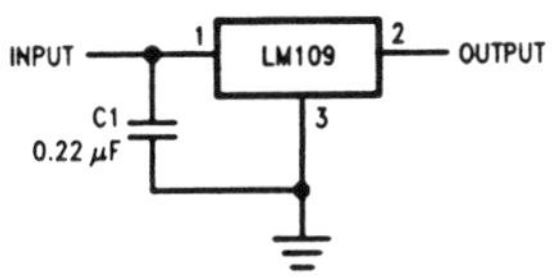

TL/H/6931-8

Figure 7–14. Characteristics and basic connections for a typical fixed three-terminal regulator (National Semiconductor, *Linear Applications Handbook,* 1991, p. 139)

7.3.1 Bypassing the Unregulated Supply

As shown in the table in Fig. 7–14, it may be necessary to bypass the unregulated input supply with a capacitor (a 0.22-μF capacitor is recommended for the LM109). The purpose of the capacitor is to prevent oscillations that can cause erratic operation. Generally, the bypass capacitor is necessary only when the regulator is located some distance from the filter capacitors on the supply output. The capacitor can generally be omitted when the regulator is used where the supply, regulator, and circuits are all on the same card.

7.3.2 A Fixed Regulator with Adjustable Output

It is possible to use most fixed regulators as an adjustable regulator for higher output voltages. Figure 7–15 shows such a circuit using the LM109. The regulated output voltage is impressed across R_1, developing a reference current. The quiescent current of the regulator, from the ground terminal, is added to the reference current. The combined currents produce a voltage drop across R_2, which raises the output voltage. Any voltage above the normal regulator output (5 V for the LM109) can be obtained as long as the voltage across the IC is kept within the ratings.

Most fixed regulators are designed so that the quiescent current is not greatly affected by variations in input voltage, load, or temperature. However, no regulator

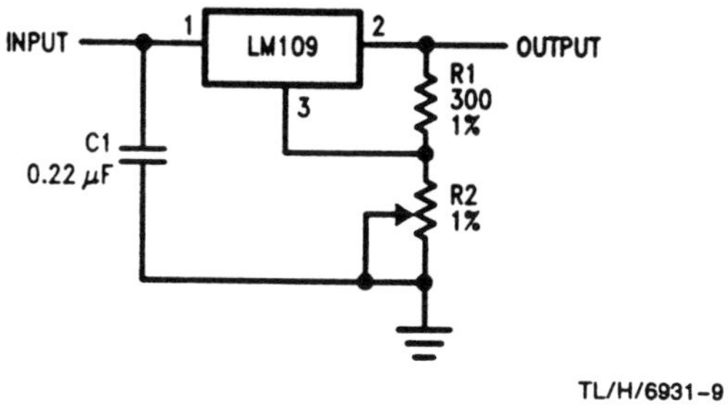

Figure 7–15. Fixed regulator with adjustable output (National Semiconductor, *Linear Applications Handbook,* 1991, p. 139)

is completely insensitive to such factors, so changes can affect regulation somewhat. For example, the quiescent current of the LM109 can vary by about 1 mA for extremes in input voltage and junction temperature. This tendency can be minimized by making the reference current through R_1 larger than the quiescent current. With the values shown in Fig. 7–15, regulation should be about 2%.

7.3.3 Fixed-Current Regulator

Figure 7–16 shows how a fixed-voltage regulator can be converted to a fixed-current regulator. The output current is set by the internal reference of the regulator and the value of R_1. (Use the information in Section 7.2.7 to calculate the value of R_1.)

7.3.4 Using a Fixed Regulator as a Pass Transistor

Because of the thermal-overload protection (where the regulator resistance increases drastically in the presence of a thermal overload), fixed regulators are considered by some as an attractive substitute for the pass transistor in feedback regulators (see Chapter 3). Figure 7–17 shows such a circuit. An op-amp compares the output voltage with a zener and controls the fixed regulator by driving the ground terminal through an FET.

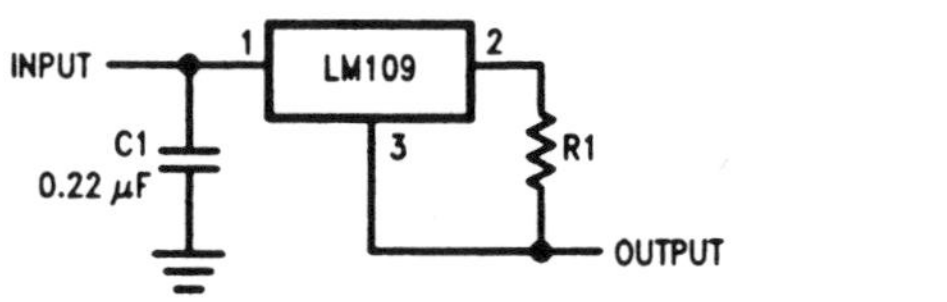

Figure 7–16. Fixed-voltage regulator converted to fixed-current regulator (National Semiconductor, *Linear Applications Handbook,* 1991, p. 140)

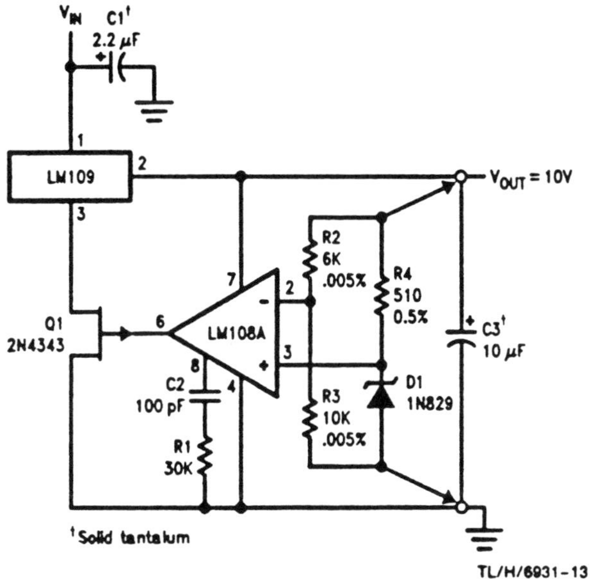

Figure 7–17. Using a fixed regulator as a series-pass transistor (National Semiconductor, *Linear Applications Handbook,* 1991, p. 140)

The load and line regulation of the Fig. 7–17 circuit is better than 0.001%. Noise, drift, and long-term stability are determined by the reference zener D_1. Noise can be further reduced by inserting 100 k 1% resistors in series with both inputs of the op-amp and bypassing the noninverting input to ground. A 100-pF capacitor can also be included between the output and the inverting input to prevent frequency instability. Temperature drift can be reduced by adjusting R_4, which determines zener current, for minimum drift. For best results, use remote sensing (see Chapter 5) directly to the load terminals.

7.4 Extending the Range of Fixed Regulators

Sometimes the need arises for a regulator voltage that is not available in standard IC regulators (or in your particular stock of regulators). If your regulator has an NPN pass transistor, the answer may be as simple as adding a resistor to the ground pin, such as R_1 in Fig. 7–18. With this circuit, the output voltage is $V_O = V_{REG} + (I_Q \times R_1)$, where V_{REG} is the standard regulator voltage and I_Q is the regulator quiescent current.

If the regulator has a PNP pass transistor, as is sometimes the case where the regulator has very low voltage drop between input and output, simply adding a resistor at the ground pin does not work. The regulator internal current varies too much because of increased base drive to compensate for lower PNP beta. However,

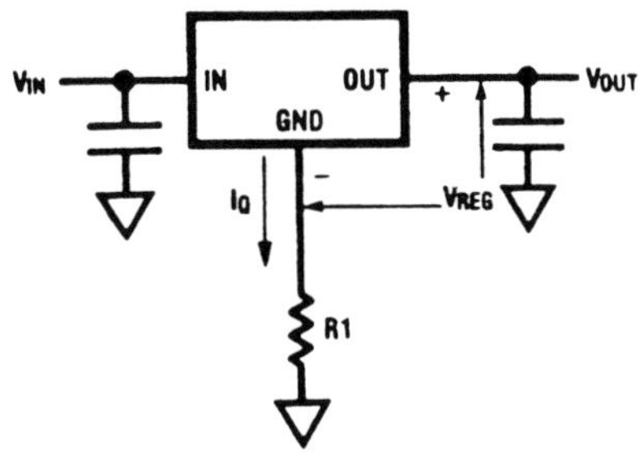

Figure 7–18. Extending range of fixed NPN regulators (National Semiconductor, *Linear Applications Handbook,* 1991, p. 1171)

if a zener is used instead of a resistor, as shown in Fig. 7–19, higher voltages can be accommodated. With this circuit, the output voltage is $V_O = V_{REG} + V_Z$, where V_Z is the zener voltage.

Since V_{REG} is constant, the output-voltage regulation depends largely on the dynamic impedance of the zener. Although zener voltage varies slightly with changes in current flow, the circuit output-voltage change is far less than with a regulator at the ground pin.

One advantage of the Fig. 7–19 circuit is the ability to achieve higher output voltages than the normal regulator rating. The maximum regulator output is limited by the breakdown of the internal circuits. However, if the zener is carefully selected to keep the input and ground-pin differential voltage well below the breakdown, the input can exceed the maximum rating. For example, a 5-V three-terminal LM2931Z PNP regulator (maximum input = 26 V) can become a 56-V regulator with a 51-V zener. The input can be as low as 56.6 V with a load current of 150 mA or less.

Most of the PNP-regulator features are still maintained with the circuit of Fig. 7–19. However, the short-circuit protection may or may not be there, depending on the output voltage and the safe operating area of the output-pass PNP transistor.

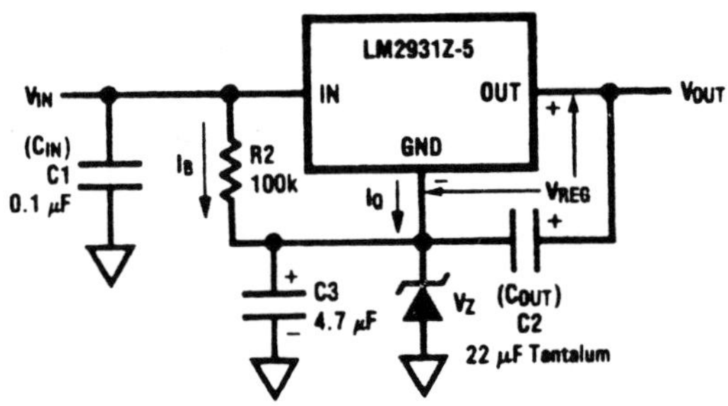

Figure 7–19. Extending range of fixed PNP regulators (National Semiconductor, *Linear Applications Handbook,* 1991, p. 1171)

Capacitors C_1 and C_2 should have the same values as those specified for normal operation. However, the maximum operating-voltage ratings should exceed the input voltage. Capacitor C_3 should be located as close as possible to the ground pin to get good decoupling and ensure stable operation. The value of C_3 depends on zener impedance and noise characteristics. The capacitor types must also be rated over the desired operating temperature range.

7.5 High-Voltage Adjustable Supplies

The floating-mode operation of adjustable three-terminal regulators makes them ideal for high-voltage operation. Since the regulator sees only input-output differential, the voltage rating will not be exceeded for outputs of hundreds of volts. However, the IC may break down when the output is shorted unless special design approaches are used to protect the IC. The following paragraphs describe two such approaches.

7.5.1 Basic High-Voltage Regulator

Figure 7–20 shows a basic high-voltage regulator using a standard adjustable regulator. Zener D_1 ensures that the LM317H sees only a 5-V input-output differen-

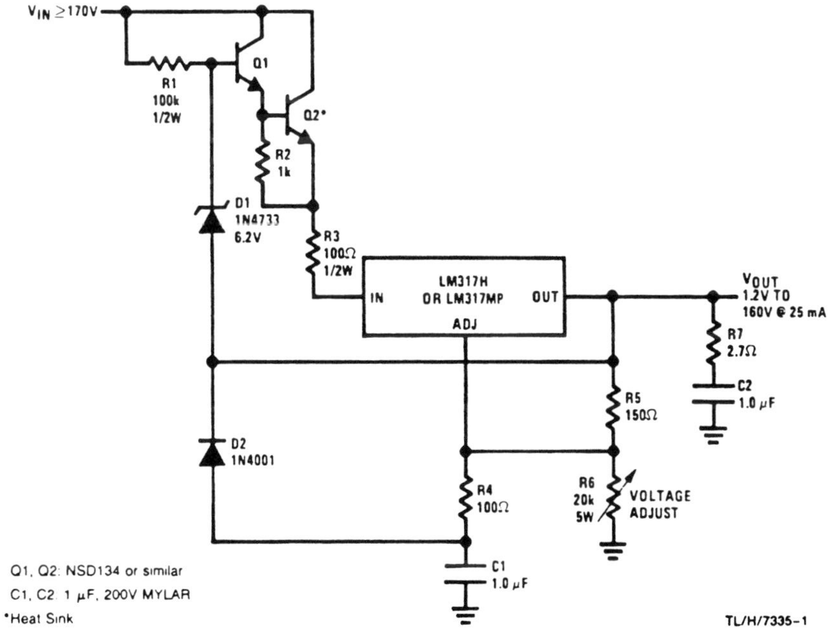

Figure 7–20. Basic high-voltage regulator (National Semiconductor, *Linear Applications Handbook,* 1991, p. 1257)

tial over the entire range of output voltage from 1.2 V to 160 V. Since high-voltage transistors by necessity have a low beta, a Darlington is used to stand off the high voltage. The zener impedance is low enough that no bypass capacitor is required at the LM317H input. (In fact, no capacitor should be used here if the circuit is to survive an output short!) R_3 limits short-circuit current to about 50 mA. The RC network on the output improves transient response, as does bypassing the adjust pin. R_4 and D_2 protect the adjust pin during shorts.

Since Q_2 may dissipate up to 5 W normally, or 10 W during a short circuit, Q_2 must have an adequate heat sink. For higher output currents, substitute a pass device (Q_1, Q_2) in a TO–3 or TO–220 package in place of the TO–202 NSD134 and reduce the value of R_3. On the other hand, if the required output current is less than 25 mA, R_3 can be increased to reduce the size of the heat sink needed for Q_1/Q_2.

7.5.2 Precision High-Voltage Regulator

Figure 7–21 shows an improved high-voltage regulator using a zener in series with the regulator internal reference. This improves both temperature stability, since the selected zener has a guaranteed TC of ±20 ppm/°C, and regulation, because more loop gain is available from the LM317.

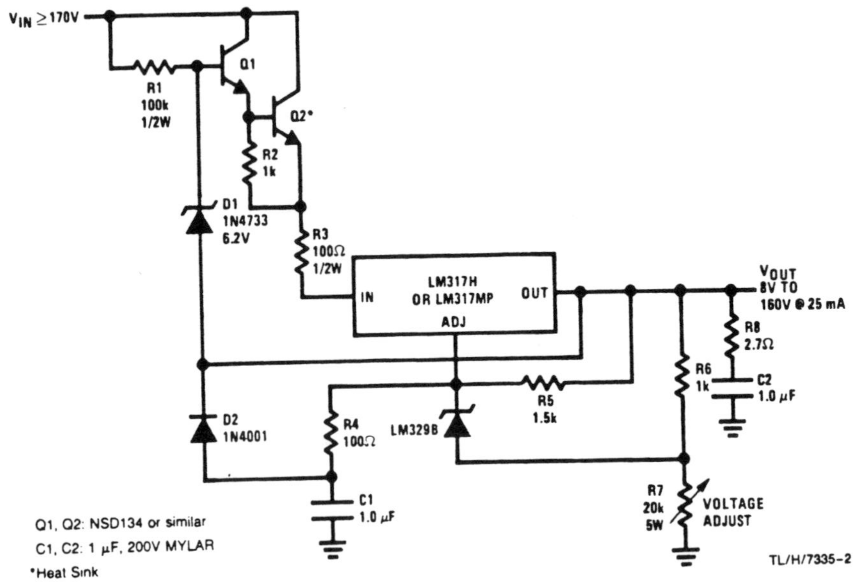

Figure 7–21. Precision high-voltage regulator (National Semiconductor, *Linear Applications Handbook,* 1991, p. 1158)

7.5.3 Extended Voltages

The techniques shown in Figs. 7–20 and 7–21 can be extended for higher output voltages and/or currents by means of better high-voltage transistors, or by cascading or paralleling several transistors (with appropriate emitter ballast resistors). If this is done, one critical point to remember is that the output short-circuit current, determined by R_3, must be within the Q_2 safe area of operation so that breakdown cannot occur.

7.6 Hybrid Voltage Regulators

Figure 7–22 shows the schematic of a typical hybrid voltage regulator (the Semtech Corpus Christi LAS 2000 series regulator). This particular line is available in many models with outputs to 28 V and 30 A. Dissipation at 25°C can reach 400 W. As shown, the hybrid has two sections, power and control. The power section contains the series-regulator power transistor (or Darlington) and a current-sensing resistor. The control section contains a voltage regulator, thick-film resistors, chip capacitors, and in some models, a tantalum output capacitor. The following paragraphs describe three design examples for the selected hybrid line.

7.6.1 Fixed Input Voltage

This example assumes that we need a power supply for use in a vehicle with a 12-V battery. The principal specifications follow:

Input voltage	10–14 V
Output voltage	5 V ± 2% (not adjustable)

THE BASIC REGULATOR
(CIRCLED NUMBERS ARE PACKAGE PIN NUMBERS)

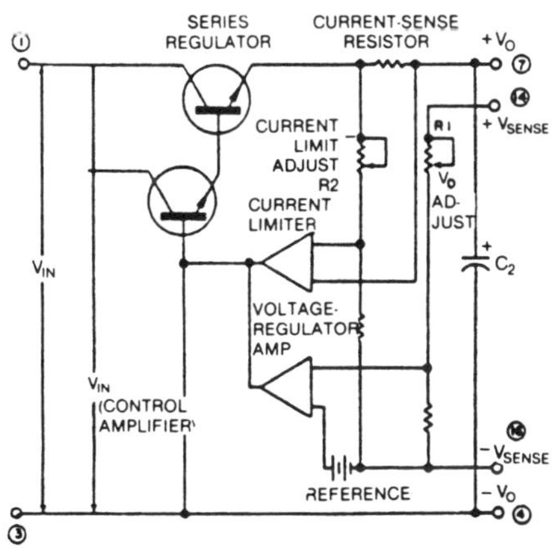

Figure 7–22. Hybrid voltage regulator (*Semtech Corporation Data Book,* 1993, p. 12–99)

Output current	0–5 A
Ambient temperature	0–55°C
Load regulation	0.2%
Line regulation	0.1%

The first step is to select a regulator that meets the basic input-output requirements. Table 7–1 shows the output ratings for some of the LAS 2000 line. Table 7–2 shows the key specifications for the line. The input range, 10 to 14 V, falls within the allowable range of 9.6 to 40 V (Table 7–2). The input-output differential, 5 to 9 V (10 – 5 to 14 – 5), is within the limits of 4.6 to 37.5 V. The output voltage, 5 V ± 2%, falls between 2.5 and 28 V. And the output current of 5 A does not exceed the maximum rating of the line. So we are now ready to select one of the regulators shown in Table 7–1.

To get 5 V at 5 A without external transistors, we could use either the LAS2105 or the LAS2205. Since there is no requirement for an adjustable output, and because our 2% tolerance requirement can be accommodated by the 1% tolerance, we choose the LAS2105.

The power dissipated in the regulator is the product of the maximum load current and the voltage differential between maximum input and minimum output. Thus, $P_{max} = 5\text{ A} \times (14 - 5)\text{ V} = 45\text{ W}$. A heat sink must be selected to allow the device to dissipate 45 W safely in an ambient of 55°C.

The graph of Fig. 7–23 shows that the maximum allowable case temperature for 45-W dissipation is 105°C. The required thermal resistance of the heat sink is $\theta_{HS} = (105 - 55)/45 = 1.1$°C/W. This calls for a rather substantial heat sink if only free-air convection and radiation are used. The heat-sink size can be reduced if forced-air cooling is used. Heat-sink dimensions are discussed further in Section 7.6.2.

Selection of a heat sink is usually based on regulator operation at the maximum rating in a nonfault mode. However, as a safety measure, it is practical to check the dissipation during a short circuit at the output. The data sheet shows that

Table 7–1 Output ratings of hybrid regulators (*Semtech Corporation Data Book,* 1993, p. 12–99)

OUTPUT RATINGS OF REPRESENTATIVE REGULATORS

Device	V_o (Volts)	I_o (Amps) (at 40°C)
LAS2105	5.0 ± 1%	5.0
LAS2205	5.0 (Trimmable)	5.0
LAS2206	6.0 (Trimmable)	5.0
LAS5205	5.0 (Trimmable)	20.0
LAS7205	5.0 (Trimmable)	30.0
LAS1505	+5.0 ± 5%	1.5
LAS1805	−5.0 ± 5%	1.5

Table 7–2 Key specifications for hybrid regulators (*Semtech Corporation Data Book,* 1993, p. 12–101)

KEY SPECIFICATIONS FOR A SERIES OF HYBRID REGULATORS — LAS 2000 SERIES

Parameter	Symbol	Minimum	Maximum	Units
Input voltage	V_{IN}	9.6	40.0	V
Output voltage[1]	V_O	2.5	28.0	V
Input-output differential[2,3]	$V_{IN} - V_O$	4.6	37.5	V
Input-output differential[3,4]	$V_{IN} - V_O$	2.5	37.5	V
Output current[1]	I_O	0	5.0	A
Standby current	I_S		10	mA
Power dissipation[5]	P_D		85	W
Power dissipation[6]	P_D		9.0	W
Thermal resistance junction—Case 1	θ_{JC}		2.0	°C/W
Thermal resistance junction—free air	θ_{JA}		15.0	°C/W
Storage temperature[7]	T_S	−55	+125	°C
Line regulation[8]			0.01	% V_{IN}
Load regulation[9]			0.2	%
Programming resistance[10]		1000		Ω/V
Temperature coefficient	T_C	0.015		%/°C
Ripple attenuation[11]		60		dB

(1) Varies with model.
(2) Single DC-input voltage.
(3) Minimum input-output differential based on T = 25°C.
(4) For separate DC-input voltages for power circuit (Pin 1) and control circuit (Pin 20), $V_{IN\ MIN}$ = 9.5V and Pin 20.
(5) Heat sink at 25°C.
(6) Free air at 25°C.
(7) Maximum storage temperature limited by tantalum capacitor.
(8) I_O constant from $V_{IN\ MIN}$ to $V_{IN\ MAX}$.
(9) V_{IN} constant from no load to full load.
(10) Nominal.
(11) Ripple attenuation (at $V_{IN\ MIN}$, $I_{O\ MAX}$) is 54 dB minimum for 20V, 24V and 28V models, 60 dB for others.

the regulator foldback-current limiting cuts the maximum current during a short circuit to 60% of the full-load rating. In this case, we have 60% of 5 A, or 3 A. So the short-circuit power is $P_{SC} = 14\ V \times 3\ A = 42\ W$. This is less than the maximum power

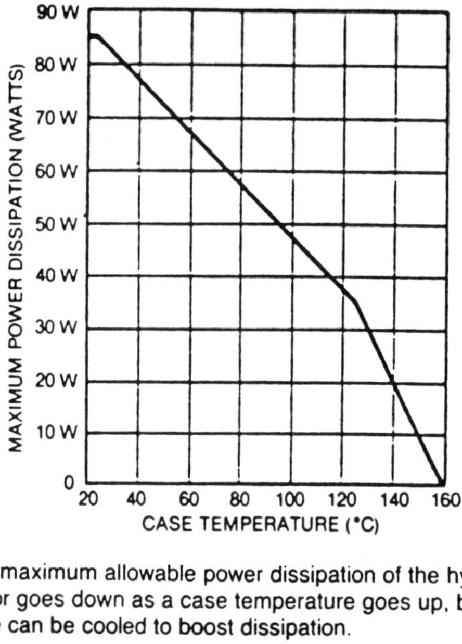

Figure 7–23. Case temperature versus maximum power dissipation of hybrid regulator (*Semtech Corporation Data Book,* 1993, p. 12–100)

under normal operating conditions (5 A), so the regulator will withstand a short circuit.

If the short-circuit power was less than 40 W, the thermal-shutdown circuit could protect the regulator by turning it off before any damage is done. In this case, there is a delay from the time the short is removed to the time the output voltage returns to within the regulation band. This is because the power section must cool down before the thermal-shutdown circuit allows the regulator to return to operation.

The final step involves checking the regulation. In this example, the requirement for load regulation is 0.2%. As shown in Table 7–2, the entire line is 0.2%. Our requirement for line regulation is 0.1%. As shown in Table 7–2, the line regulation is 0.01% per volt of line change. The maximum line change in our example is 4 V (10 to 14 V), so the maximum line regulation is 0.04%. This is well within our 0.1% requirement.

These design figures show that the LAS2105 meets all of our requirements, as is. No external components are needed, and design is complete.

7.6.2 Wide-Range AC Input

In this example, the input is variable and we must specify the proper transformer secondary voltages, hybrid regulator, and heat sink. The key requirements are as follows:

Input	To be specified, based on line variation of 105–132 V at 60 Hz
Output voltage	6 V ± 5% (trimmable)
Output current	0–2.8 A
Ambient temperature	0–40°C
Load regulation	0.2%
Line regulation	0.2%
Output ripple	5 mV p-p (maximum)

As shown in Table 7–2, the minimum input voltage is 9.6 V. However, Table 7–2 also requires a minimum input-output differential of 4.6 V. Since our output is 6 V ± 5%, the output could go to 6.3 V (1.05 × 6). Add this to 4.6 V for a minimum instantaneous voltage on the input filter capacitor of 10.9 V (at low line and full load).

To determine the minimum filter capacitance, consider the output-ripple requirement of 5 mV p-p. Since the ripple attenuation of the regulator is 60 dB, or times 1000, as shown in Table 7–2, the maximum ripple that can appear on the capacitor is 5 V p-p. Semtech recommends 1000 μF/A for 60 Hz as a minimum filter capacitance. As a result, this requires a 2800-μF filter capacitor for our 2.8-A maximum output current.

Notice that the 1000-μF/A figure is a guideline, suitable as a first trial value for simplified design. The exact ripple is a function of the transformer source impedance, the capacitor, and the load current. The transformer specifications depend in part on the specific circuit we use external to the regulator. Let us consider the design shown in Fig. 7–24. An alternate design is discussed in Section 7.6.3.

If the transformer-capacitor combination is designed to yield a maximum of 5 V p-p ripple at high line, the ripple at low line will be somewhat less, usually about 4.5 V. The average voltage on the capacitor at low line is the sum of the minimum V_{IN} value and half the peak-to-peak ripple, or V_{IN} (average) = 10.9 + 2.25 = 13.15 V.

The maximum power dissipated depends on the input voltage at high line, which is V_{IN} (average) = (132/105) × 1.1 × 13.15 = 18.2 V, where the 1.1 factor ac-

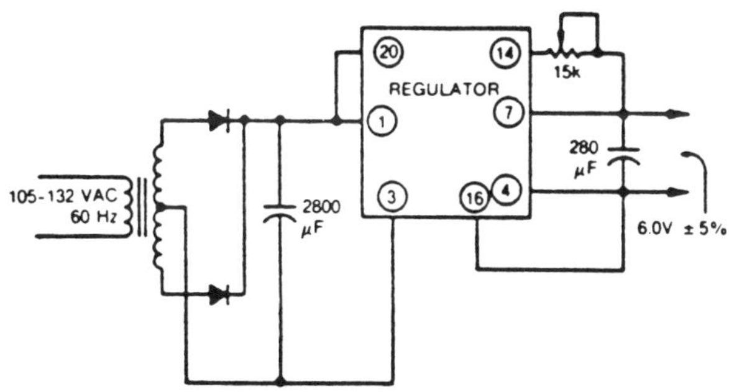

Figure 7–24. Hybrid regulator for wide-range AC input (*Semtech Corporation Data Book,* 1993, p. 12–102)

counts for the change in rectification efficiency that occurs during the transition from low (105 V) to high (132 V) line. In this case, the transformer must provide 51 W (18.2 V at 2.8 A).

Now that the transformer specifications are established, the normal operating power dissipation of the regulator must be found. Regulator dissipation is P = (18.2 – 5.7 V) × 2.8 A = 35 W. The 5.7-V value is the output when adjusted to 5% below the nominal 6.0-V level. This low-output value must be used since the regulator dissipation depends on the maximum input-output differential.

As shown in Fig. 7–23, the maximum allowable case temperature with a dissipation of 35 W is 125°C. The required thermal resistance of the heat sink is θ_{HS} = (125 – 40)/35 = 2.4°C/W. As shown in Fig. 7–25, a horizontal heat sink measuring 5 × 5 × 3/32 inches would dissipate about 28 W. Mounting a pointed heat sink vertically should increase dissipation to about 35 W.

As shown in Table 7–1, to get an adjustable or trimmable 6 V at 2.8 A, we must choose the LAS2206. As discussed in Section 7.6.1, the maximum short-circuit current is 60% of the device (LAS2206) full-load current, or 5 A (not our 2.8-A output current). So the short-circuit power is P_{SC} = 18.2 V × 3 A = 54.6 W. As a safety measure, use a heat sink that will accommodate the short-circuit power rather than the normal operating power.

The next step involves checking the regulation. In this example, the requirement for load regulation is 0.2%. As shown in Table 7–2, the entire LAS 2000 line is 0.2%. Our requirement for line regulation is also 0.2%. As shown in Table 7–2,

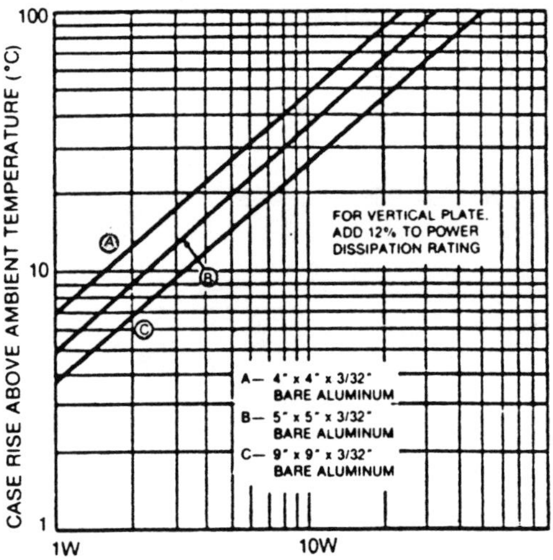

ANOTHER 12% CAN BE ADDED TO THE POWER DISSIPATION RATING FOR A VERTICAL PLATE AND AN ADDITIONAL 10% CAN BE ADDED IF SURFACES ARE PAINTED.

Figure 7–25. Typical heat-sink data for horizontal plate (*Semtech Corporation Data Book,* 1992, p. 12–103)

the line regulation is 0.01% per volt of line change. The change in voltage across the filter capacitor when the input line is varied from 105 to 132 V is $\Delta V = 18.2 - 13.15 = 5.05$ V. So the maximum line regulation is 0.01%/V × 5.05 V = 0.05%. This is well within our 0.2% requirement.

The final step involves determining the values for the voltage control (trim control) and for the output capacitor. The manufacturer recommends a 15-k potentiometer for the voltage control (between pins 7 and 14, as shown in Fig. 7–24). The potentiometer should be selected for high stability and low temperature coefficient because any variation results in an output-voltage change.

The output capacitor should be at least 100 μF/A, or 280 μF in our case. Use a high-grade aluminum or tantalum electrolytic with low ESR (equivalent series resistance). The capacitor voltage rating should be such that if an external sense lead opens and the output loses regulation, the capacitor rating will not be exceeded by the unregulated output voltage.

7.6.3 Wide-Range AC Input (Alternate Design)

Figure 7–26 shows an alternate circuit to that given in Fig. 7–24. The alternate circuit requires more components but results in a smaller transformer and heat sink. The key requirements remain the same. In the circuit of Fig. 7–26, the control input (pin 20) and the power input (pin 1) are separated. As shown in Table 7–2, the power circuit requires that the voltage differential between pin 1 and the output (pin 7) be at least 2.5 V. Therefore, $V_{IN} = (1.05 \times 6) + 2.5\ V = 8.8$ V at pin 1.

If the same capacitor used in Section 7.6.2 is still employed, the average input voltage at low line is $V_{IN} = 8.8 + 2.25\ V = 11.05$ V at 105 V line. Similarly, $V_{IN} = 1.38 \times 11.05 = 15.2$ V at 132 V line. The maximum power dissipation is then $P = (15.2 - 5.7) \times 2.8\ A = 26.6$ W. This is a reduction of about 25%. The transformer output is reduced to 42.5 W (15.2 V at 2.8 A), a saving of 17%.

Now let us consider the control circuit. The minimum differential between pin 20 and pin 7 (which is at 6 V ± 5%) is 4.6 V. Thus, pin 20 must never see less than

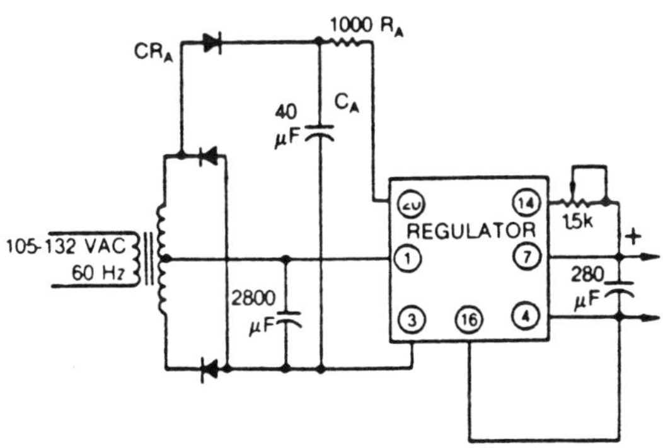

Figure 7–26. Alternate hybrid regulator for wide-range AC input (*Semtech Corporation Data Book,* 1992, p. 12–103)

10.9 V. In Fig. 7–26, the maximum voltage of the peak-detector capacitor C_A is twice the highest voltage on the main filter capacitor. Thus, at low-line, full-load conditions, the peak voltage on the main capacitor is 13.3 V and the voltage on C_A is about 26.2 V. This is more than adequate biasing (far above 10.9 V) for the control input at pin 20.

The next step is to calculate the value for the peak-detector capacitor C_A and resistor R_A (which limits the current into control pin 20 during turnoff). The capacitor value depends primarily on ripple. The ripple attenuation of the regulator is defined as the ratio of the ripple at the control-circuit input to that at the output. To maintain no more than 5 mV at the output, the maximum ripple on capacitor C_A must be 5 V p-p. Capacitor C_A is connected in a half-wave configuration with control diode CR_A. (Half wave is used to minimize the number of components.)

In a half-wave 60-Hz system, the conduction time of diode CR_A is about 3 ms. The discharge time of capacitor C_A is therefore 13.6 ms (the line period of 16.6 ms minus 3 ms). Table 7–2 shows that the maximum input standby current to pin 20 is 10 mA. The minimum value of capacitor C_A is the product of the standby current and the discharge time, divided by the maximum ripple voltage, or C_{min} = 10 mA × (13.6 ms/5 V) = 27.2 μF. The manufacturer recommends that 40 μF be used for C_A to provide a safety margin.

Notice that the circuit for the peak-detector/control-input (diode CR_A and capacitor C_A) is essentially a voltage doubler, as discussed in Chapter 2. As a result, the voltage at high-line, full-load conditions is about 2 × (15.2 + 2.5 V) = 35.4 V. This is close to the maximum input of 40 V and limits the Fig. 7–26 circuit to applications where the output voltage is less than about 7 or 8 V, depending on the range of the input-line swing.

We know that the load regulation for the entire LAS 2000 line is 0.2% (Table 7–2), so we need not calculate load regulation for the Fig. 7–26 circuit. Table 7–2 also shows that the line regulation is 0.01% per volt of line change. In the alternate design of Fig. 7–26, the change sensed by the control input is 8.8 V (35.4 – 26.6 V). So the maximum line regulation is 0.01%/V × 8.8 V, or about 0.09%. This is well within our 0.2% requirement.

The next component that must be considered is current-limit resistor R_A. During turnoff, the regulator tries to provide load current from the peak-detector/control circuit. The control circuit cannot carry much current, even for a very short time. The control circuit (and thus the entire regulator) will be destroyed if the current is not limited to 150 mA. The following calculations are used to find a suitable value for R_A.

When the power supply is turned off, both the main capacitor and capacitor C_A start to discharge. When the main capacitor discharges below the voltage required at pin 1 to maintain the output, the load current is drawn from the peak-detector/control circuit.

Start by assuming that the power supply is operating at high-line, full-load capacity and then is turned off. Calculate the time required for the main capacitor to discharge to the minimum voltage at pin 1. Then calculate the value to which the voltage on C_A has decayed in the same time. This voltage appears across R_A and is used to find the minimum value for R_A.

To simplify calculations, assume that the following voltages are present at turnoff time.

Capacitor C_A = 36 V
Main capacitor = 17.7 V

When the main capacitor discharges to 8.5 V (6.0 V output + 2.5 V minimum input-output differential), C_A starts to supply the current. The time required for this to happen is

$$t = CV/I$$

$$t = [2800\ \mu F\ (17.7 - 8.5\ V)]/2.8\ A = 9.2\ ms$$

During this time, the voltage change on C_A is

$$\Delta V = It/C$$

$$= (10\ mA \times 9.2\ ms/40\ \mu F = 2.3\ V$$

The voltage remaining on C_A is 33.7 V (36 – 2.3). This voltage appears across R_A, and requires a resistance of at least 225 ohms to limit the current to 150 mA. There is also a limitation on the maximum value of R_A. In normal operation, the standby current (see Table 7–2) is 10 mA, which results in a voltage drop across R_A. The value of R_A must be selected so that under low-line, full-load, and maximum-output-voltage conditions, there is enough voltage at pin 20 for proper operation. As discussed, pin 20 must never see less than 10.9 V.

The minimum instantaneous voltage on C_A is the low-line peak of 26.6 V, minus the peak-to-peak ripple (which in turn depends on capacitance value). As discussed, a value of 27.2 µF should produce a 5-V ripple, so a 40-µF value should produce an approximate 3.5-V ripple (40/5 = 27.2/3.5). Thus, the minimum instantaneous voltage on C_A is (26.6 – 3.5), or 23.1 V. Using this voltage and the 10-mA standby current, the maximum value of R_A is (23.1 – 10.9)/10 mA = 1220 ohms. The range for R_A is then somewhere between 225 and 1220 ohms. The recommended Fig. 7–26 value is 1000 ohms.

The criteria for selection of the voltage or trim control and the output capacitor are the same as described in Section 7.6.2. However, notice that the circuit of Fig. 7–26 reduces transformer size by about 15 to 20%. The alternate design also reduces heat-sink size by 25%, all for the price of three extra components.

7.7 Programmable Voltage Regulators

Figure 7–27 shows the pinouts and functional block diagram for typical programmable regulators (the Harris Semiconductor CA723/CA723C). Table 7–3 shows the key design characteristics. Such regulators provide much better regulation

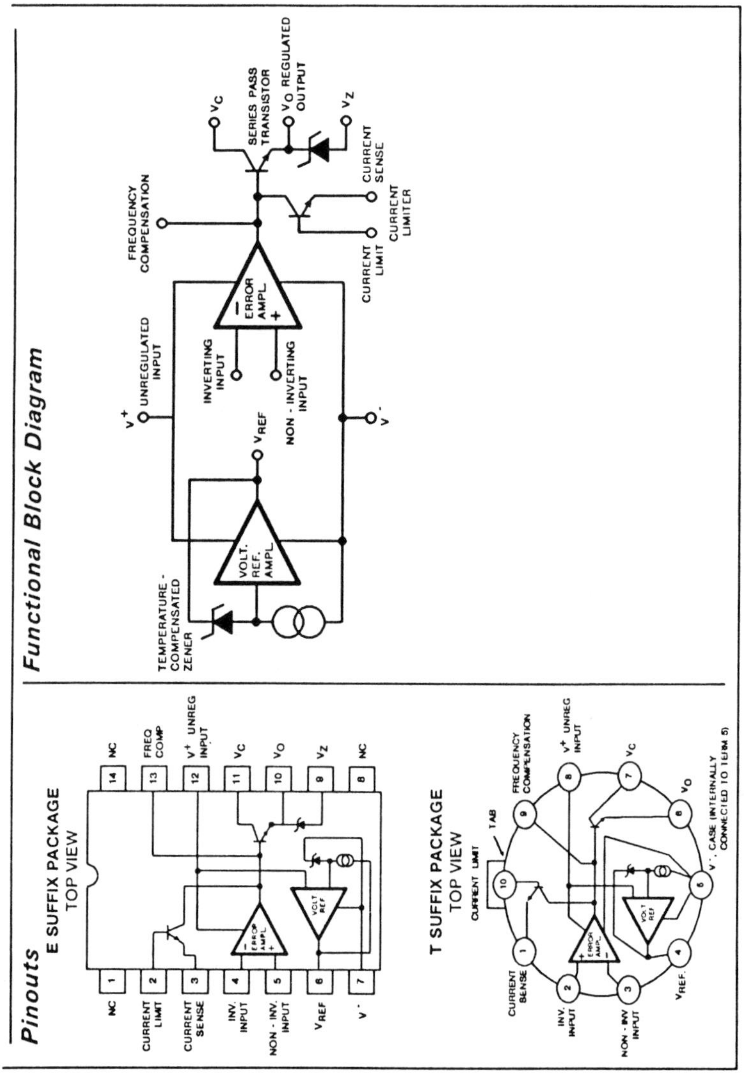

Figure 7-27. Pinouts and functional diagram for CA723/723C programmable voltage regulator (Harris Semiconductor, *Linear & Telecom ICs*, 1991, p. 2–5)

Table 7–3. Key design characteristics for CA723/723C regulators (Harris Semiconductor, *Linear & Telecom ICs,* 1991, p. 2–7)

ELECTRICAL CHARACTERISTICS at T_A = 25 C, $V^+ = V_C = V_I$ = 12 V, V^- = 0, V_O = 5 V, I_L = 1 mA, C_1 = 100 pF, C_{REF} = 0, R_{SCP} = 0, unless otherwise specified. Divider impedance $\frac{R_1 R_2}{R_1+R_2}$ at non-inverting input, Term. 5, = 10 kΩ (See Figure 20)

CHARACTERISTIC	TEST CONDITIONS	LIMITS CA723 Min.	Typ.	Max.	LIMITS CA723C Min.	Typ.	Max.	UNITS
Quiescent Regulator Current, I_Q	I_L = 0, V_I = 30 V	–	2.3	3.5	–	2.3	4	mA
Input Voltage Range, V_I		9.5	–	40	9.5	–	40	V
Output Voltage Range, V_O		2	–	37	2	–	37	V
Differential Input-Output Voltage, $V_I - V_O$		3	–	38	3	–	38	V
Reference Voltage, V_{REF}		6.95	7.15	7.35	6.8	7.15	7.5	V
Line Regulation (See Note 1)	V_I = 12 to 40 V	–	0.02	0.2	–	0.1	0.5	$\%V_O$
	V_I = 12 to 15 V	–	0.01	0.1	–	0.01	0.1	
	V_I = 12 to 15 V, T_A = −55 to +125°C	–	–	0.3	–	–	–	
	V_I = 12 to 15 V, T_A = 0 to 70°C	–	–	–	–	–	0.3	
Load Regulation (See Note 1)	I_L = 1 to 50 mA	–	0.03	0.15	–	0.03	0.2	$\%V_O$
	I_L = 1 to 50 mA, T_A = −55 to +125°C	–	–	0.6	–	–	–	
	I_L = 1 to 50 mA, T_A = 0 to 70°C	–	–	–	–	–	0.6	
Output-Voltage Temp. Coefficient, ΔV_O	T_A = −55 to +125°C	–	0.002	0.015	–	–	–	%/°C
	T_A = 0 to 70°C	–	–	–	–	0.003	0.015	
Ripple Rejection (See Note 2)	f = 50 Hz to 10 kHz	–	74	–	–	74	–	dB
	f = 50 Hz to 10 kHz, C_{REF} = 5 μF	–	86	–	–	86	–	
Short-Circuit Limiting Current, I_{LIM}	R_{SCP} = 10 Ω, V_O = 0	–	65	–	–	65	–	mA
Equivalent Noise RMS Output Voltage, V_N (See Note 2)	BW = 100 Hz to 10 kHz, C_{REF} = 0	–	20	–	–	20	–	μV
	BW = 100 Hz 10 kHz, C_{REF} = 5 μF	–	2.5	–	–	2.5	–	

Note 1: Line and load regulation specifications are given for condition of a constant chip temperature. For high-dissipation condition, temperature drifts must be separately taken into account.

Note 2: For C_{REF} (See Figure 20).

than the three-terminal or hybrid devices discussed thus far in this chapter. As shown in Table 7–3, typical line and load regulation is in the 0.01 to 0.03% range.

These particular regulators provide output voltages ranging from 2 to 37 V at currents up to 150 mA. As shown in Fig. 7–27, the regulators contain a temperature-compensated reference amplifier, a power series-pass transistor, and a current-limiting circuit. The devices also provide independently accessible inputs for adjustable current limiting and remote shutdown. Standby-current drain is low (typically 2 to 3 mA), as is temperature drift (typically 0.002%/°C). Ripple rejection is high (typically 74 to 86 dB) and it can be increased when an external capacitor is used.

These regulators may be used with positive and negative power supplies in a wide variety of series, shunt, switching, and floating regulator applications. The devices can provide regulation at load currents greater than 150 mA—and in excess of 10 A with suitable external pass transistors (both NPN and PNP). The regulators are supplied in the 10-lead, TO–5-style package (T suffix) and the 14-lead, DIP plastic package (E suffix). The devices are direct replacements for industry types 723, 723C, μA723 and μA723C, in packages with similar terminal arrangements. The regulators are also available in chip form (H suffix) and are rated for operation over the full military-temperature range of –55°C to +125°C. The following paragraphs describe some design examples for the line.

7.7.1 Low-Voltage Regulator

Figure 7–28 shows the CA723 connected to provide regulation at voltages between 2 and 7 V. As usual, the output voltage depends on the ratio R_1/R_2. As shown

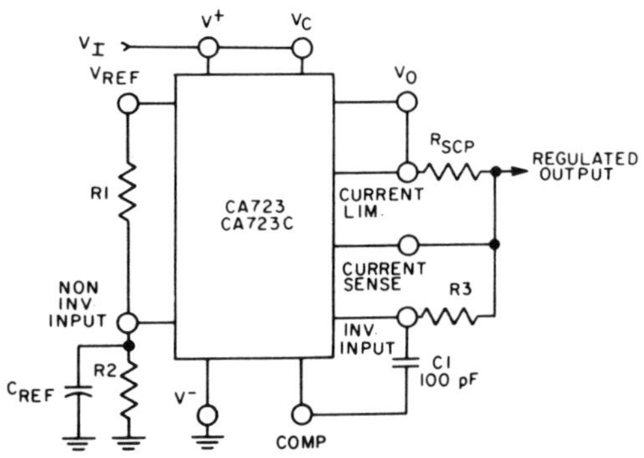

CIRCUIT PERFORMANCE DATA:

REGULATED OUTPUT VOLTAGE . . .	5	V
LINE REGULATION (ΔV_I = 3 V)	0.5	mV
LOAD REGULATION (ΔI_L = 50 mA). . .	1.5	mV

Note: $R3 = \frac{R1\ R2}{R1+R2}$ for minimum temperature drift

Figure 7–28. Low-voltage regulator (Harris Semiconductor, *Linear & Telecom ICs,* 1991, p. 2–11)

in Table 7–3, V_{REF} is about 7 V. So for an output of 5 V, R_1 should drop the reference by about 2 V, while R_2 drops the remaining 5 V to ground, and the noninverting input is at 5 V. Under no circumstance should the values of R_1/R_2 be such that the current from V_{REF} is above 15 mA. As a first trial value, select R_1 and R_2 such that 1 mA is drawn from V_{REF}. For example, with V_{REF} at 7 V, the total values of $R_1 + R_2$ should be 7 k (R_1 = 2 k and R_2 = 5 k, for a 5-V output).

The trial value of R_3 should be equal to the actual divider impedance at the noninverting input. This condition produces minimum temperature drift.

The value of R_{SCP} should be such that the voltage drop across it is about 0.5 to 0.7 V when the desired current-limit point is reached. For example, to limit the current at 100 mA, use a value of 5 to 7 ohms for R_{SCP}, as a first trial value. This should be sufficient to turn off the internal series-pass transistor. As shown in Fig. 7–27, the series-pass transistor is turned off when the current-limiter transistor is turned on by sufficient base-emitter differential (or the voltage across R_{SCP}).

The value of frequency-compensation capacitor C_1 is set at 100 pF, as shown in Fig. 7–28. The value of the C_{REF} capacitor depends on the amount of ripple rejection and noise voltage desired. As shown in Table 7–3, typical ripple rejection is 74 dB and equivalent noise is 20 μV when C_{REF} is omitted. With a C_{REF} of 5 μF, ripple rejection increases to 86 dB and noise drops to 2.5 μV.

For simplified design, consider the total device dissipation to be the product of the input-output voltage differential and the output current. (The other factors of quiescent current, V_{REF} current, etc., can generally be ignored.) Although, in theory, the devices can dissipate up to about 1 W at 25°C, use 750 mW. Above 25°C, derate linearly at about 7 mW/°C.

If the full 150-mA output is used, then the maximum input-output differential is limited to 5 V (0.75/5.0 = 0.150). The data-sheet graphs show that the devices will deliver the full 150 mA at 25°C with an input-output differential of 5 V. The maximum output drops off to 100 mA with a differential of 8 V, and to 60 mA when the differential increases to the absolute maximum of 40 V. So always design for a minimum input-output differential, if practical.

7.7.2 High-Voltage Regulator

Figure 7–29 shows the CA723 connected to provide regulation at voltages between 7 and 37 V. Again, the output voltage depends on the ratio R_1/R_2. However, in this case R_1 and R_2 monitor the output voltage, with the R_1/R_2 junction connected to the inverting input. This multiplies the V_{REF} connected to the noninverting input through R_3. For example, with both R_1 and R_2 at the same approximate value (say, 7 k for R_2 and 8 k for R_1), the voltage at the two inputs will be about 7 V, but the regulated output voltage will be 15 V.

All of the design considerations discussed in Section 7.7.1 apply here, except that C_{REF} is omitted and R_3 may be omitted if a minimum component count is essential.

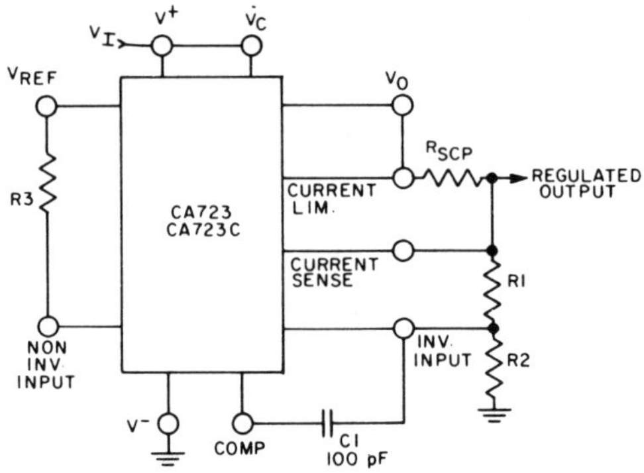

CIRCUIT PERFORMANCE DATA:

REGULATED OUTPUT VOLTAGE . . .	15	V
LINE REGULATION (ΔV_I = 3 V)	1.5	mV
LOAD REGULATION (ΔI_L = 50 mA) . .	4.5	mV

Note: $R3 = \frac{R1\ R2}{R1+R2}$ for minimum temperature drift

R3 may be eliminated for minimum component count.

Figure 7–29. High-voltage regulator (Harris Semiconductor, *Linear & Telecom ICs,* 1991, p. 2–11)

7.7.3 Negative Voltage Regulator

Figure 7–30 shows the CA723 connected to provide regulation for negative voltages at currents limited only by the external pass transistor (a PNP because of

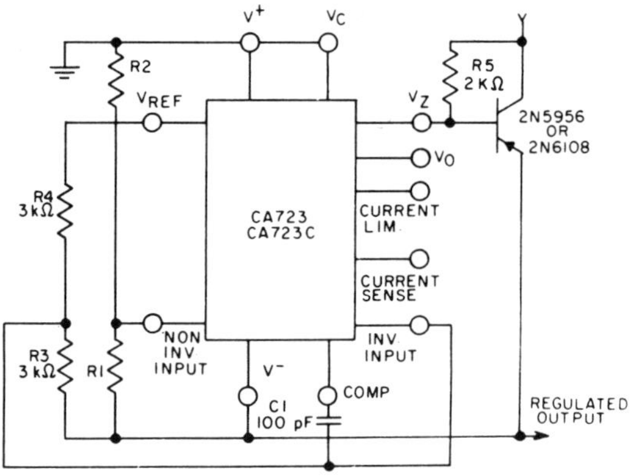

CIRCUIT PERFORMANCE DATA:

REGULATED OUTPUT VOLTAGE . . .	−15	V
LINE REGULATION (ΔV_I = 3 V)	1	mV
LOAD REGULATION (ΔI_L = 100 mA) . .	2	mV

Note: For applications employing the TO-5 style package and where V_Z is required, an external 6.2-volt zener diode should be connected in series with V_O (Terminal 6).

Figure 7–30. Negative voltage regulator (Harris Semiconductor, *Linear & Telecom ICs,* 1991, p. 2–11)

the negative voltage). Here the inverting input receives one-half the differential between V_{REF} and output (because R_3 and R_4 are the same value). The values of R_1 and R_2 should be such that the noninverting input is equal to that of the inverting input. The design considerations discussed in Section 7.7.1 apply. Make certain to observe the note in Fig. 7–30 concerning the TO–5 package.

7.7.4 Positive Voltage Regulator (with External NPN)

Figure 7–31 shows the CA723 connected to provide regulation for positive voltages at currents limited only by the external pass transistor. Here the noninverting input receives the full V_{REF} (about 7 V). The values of R_1 and R_2 should be such that the inverting input is equal to that of the noninverting input. The drop across R_2 should be about 7 V. Try 8 k for R_1 and 7 k for R_2 to get 15 V output. The design considerations discussed in Section 7.7.1 apply, except that the value of frequency-compensation capacitor C_1 is 500 pF.

7.7.5 Positive Voltage Regulator (with External PNP)

Figure 7–32 shows the CA723 connected to provide regulation for positive voltages at currents limited only by the external pass transistor. Here the inverting input is connected directly to the output voltage (5 V in the example shown). The values of R_1 and R_2 should be such that the noninverting input is also at 5 V. Try 2 k

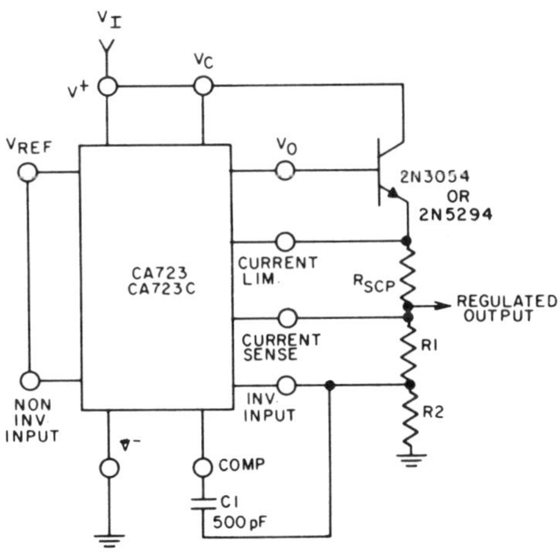

CIRCUIT PERFORMANCE DATA:

REGULATED OUTPUT VOLTAGE . . .	15 V
LINE REGULATION (ΔV_I = 3 V)	1.5 mV
LOAD REGULATION (ΔI_L = 1 A)	15 mV

Figure 7–31. Positive voltage regulator (with external NPN)(Harris Semiconductor, *Linear & Telecom ICs,* 1991, p. 2–11)

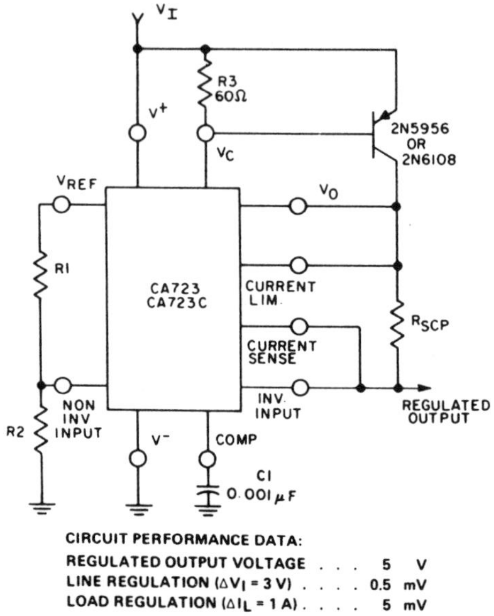

Figure 7–32. Positive voltage regulator (with external PNP)(Harris Semiconductor, *Linear & Telecom ICs,* 1991, pp. 2–12)

for R_1 and 5 k for R_2 to get 5 V output. The design considerations discussed in Section 7.7.1 apply, except that the value of capacitor C_1 is 0.001 μF.

7.7.6 Regulator with Internal Current Foldback

Figure 7–33 shows the CA723 connected to provide current foldback in case of a short circuit. This is similar to the current foldback described in Chapter 4. However, the circuit of Fig. 7–33 does not require an external transistor for short-circuit protection. Instead, the base of the internal current-limiter transistor (see Fig. 7–27) is connected to the junction of R_3/R_4, while the emitter is connected directly to the output voltage. In case of a short circuit at the output, the output current drops to 20 mA. Since the inverting input is connected directly to the output voltage (5 V in the example shown), the values of R_1 and R_2 should be such that the noninverting input is also at 5 V. Try 2 k for R_1 and 5 k for R_2. The design considerations discussed in Section 7.7.1 apply, except that the value of capacitor C_1 is 0.001 μF, and the first trial value for R_{SCP} should be 30 ohms.

7.7.7 Positive Floating Regulator

Figure 7–34 shows the CA723 used to stabilize voltages considerably higher than the ratings of the IC. In this case, an 85-V source is regulated at 50 V. This is similar to the floating op-amp regulator described in Chapter 5. However, the circuit

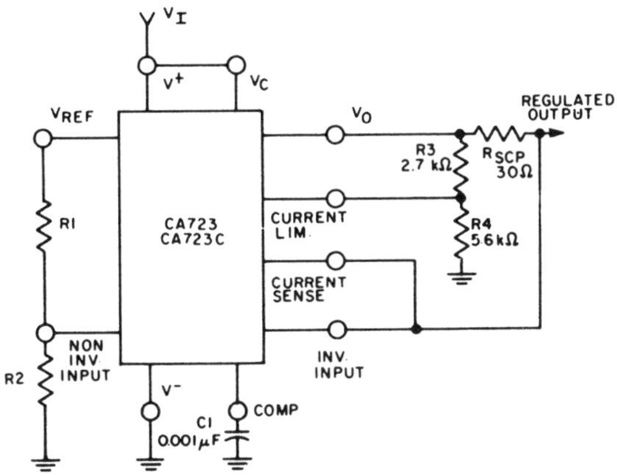

CIRCUIT PERFORMANCE DATA:

REGULATED OUTPUT VOLTAGE . . . 5 V
LINE REGULATION (ΔV_I = 3 V) . . . . 0.5 mV
LOAD REGULATION (ΔI_L = 10 mA). . . 1 mV
SHORT-CIRCUIT CURRENT . . . . . . 20 mA

Figure 7–33. Regulator with internal current foldback (Harris Semiconductor, *Linear & Telecom ICs,* 1991, p. 2–12)

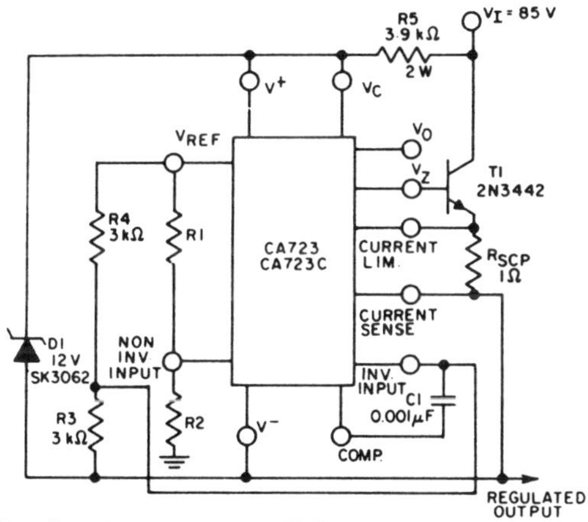

Note: For applications employing the TO-5 style package and where V_Z is required, an external 6.2-volt zener diode should be connected in series with V_O (Terminal 6).

CIRCUIT PERFORMANCE DATA:

REGULATED OUTPUT VOLTAGE . . . 50 V
LINE REGULATION (ΔV_I = 20 V) . . . 15 mV
LOAD REGULATION (ΔI_L = 50 mA). . . 20 mV

Figure 7–34. Positive floating regulator (Harris Semiconductor, *Linear & Telecom ICs,* 1991, p. 2–12)

of Fig. 7–34 uses an IC regulator rather than an op-amp. Operation of this circuit is possible because the entire circuit is not referenced to ground but is "floated" between ground and the supply voltage. Notice that the differential between the V^+ and V^- terminals of the CA723 is set by D_1 (12 V in this case). Again, the regulated output is set by the values of R_1 and R_2. (This sets the value at the noninverting input.) The voltage at the inverting input is one-half the differential between V_{REF} and the regulated output. The design considerations discussed in Section 7.7.1 apply, except that the value of capacitor C_1 is 0.001 μF and the first trial value of R_{SCP} should be 1 ohm.

7.7.8 Negative Floating Regulator

Figure 7–35 shows the CA723 used to stabilize a negative voltage considerably higher than the ratings of the IC. In this case, the output is regulated at –100 V. Again, this is similar to the floating op-amp regulator described in Chapter 5, except that the circuit uses an IC regulator rather than an op-amp. Since a negative voltage is controlled, the pass transistor T_1 is PNP rather than NPN. As in the case of the Fig. 7–34 circuit, the differential between the V^+ and V^- terminals of the CA723 is set by D_1 (12 V), and the regulated output is set by the values of R_1 and R_2. Notice that the current-limit function is not used, and there is no R_{SCP} resistor.

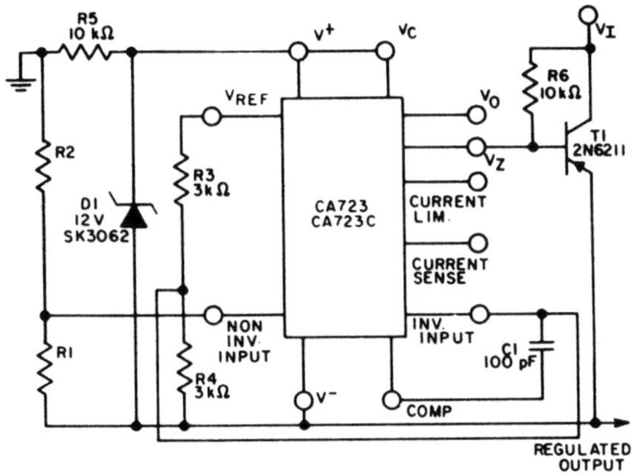

CIRCUIT PERFORMANCE DATA:

REGULATED OUTPUT VOLTAGE . . .	–100	V
LINE REGULATION (ΔV_I = 20 V) . . .	30	mV
LOAD REGULATION (ΔI_L = 100 mA) . .	20	mV

Note: For applications employing the TO-5 style package and where V_Z is required, an external 6.2-volt zener diode should be connected in series with V_O (Terminal 6).

Figure 7–35. Negative floating regulator (Harris Semiconductor, *Linear & Telecom ICs,* 1991, p. 2–12)

7.7.9 Remote-Shutdown Regulator with Current Limiting

Figure 7–36 shows the CA723 used to provide both external shutdown by means of a logic signal, and current limiting. This is similar to the current-limiting and shutdown circuits described in Chapter 4. Since the inverting input is connected directly to the output voltage (5 V in the example shown), the values of R_1 and R_2 should be such that the non-inverting input is also at 5 V. The design considerations discussed in Section 7.7.1 apply, except that the value of capacitor C_1 is 0.001 μF.

7.7.10 Shunt Regulator

Figure 7–37 shows the CA723 used as the control element in a shunt regulator. This is similar to the shunt regulators described in Chapter 3, except that the CA723 is used to control shunt transistor T_1 instead of a transistor-zener combination. Here the non-inverting input is connected directly to the output voltage (5 V in the example shown), and the values of R_1 and R_2 must be selected to set the inverting input at 5 V (or at the desired output voltage). Notice that the current-limiting function is not used, and there is no R_{SCP} resistor. The design considerations discussed in Section 7.7.1 apply, except that the value of capacitor C_1 is 0.005 μF.

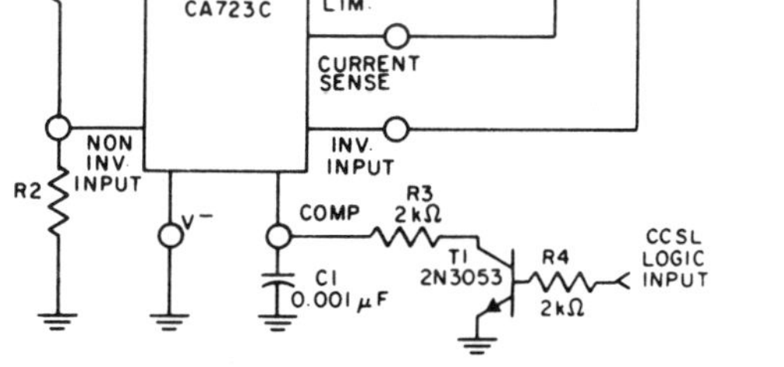

CIRCUIT PERFORMANCE DATA:
REGULATED OUTPUT VOLTAGE . . . 5 V
LINE REGULATION (ΔV_I = 3 V) . . . . 0.5 mV
LOAD REGULATION (ΔI_L = 50 mA). . . 1.5 mV

Note 1: A current limiting transistor may be used for shutdown if current limiting is not required.
Note 2: Add a diode if V_O > 10 V.

Figure 7–36. Remote shutdown regulator with current limiting (Harris Semiconductor, *Linear & Telecom ICs,* 1991, p. 2–12)

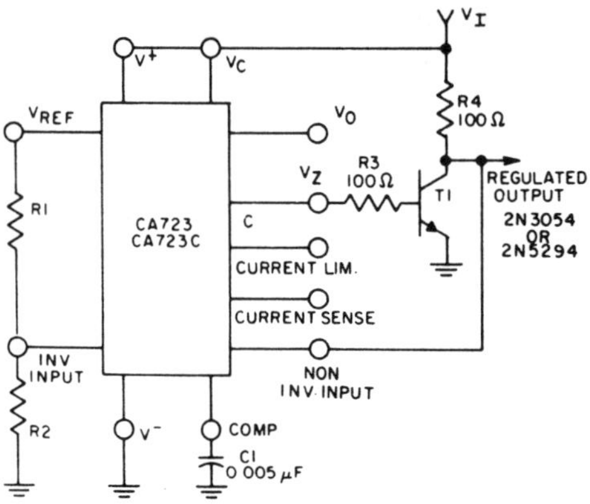

CIRCUIT PERFORMANCE DATA:
REGULATED OUTPUT VOLTAGE . . . 5 V
LINE REGULATION (ΔV_I = 10 V) . . . 0.5 mV
LOAD REGULATION (ΔI_L = 100 mA) . . 1.5 mV

Note: For applications employing the TO-5 style package and where V_Z is required, an external 6.2-volt zener diode should be connected in series with V_O (Terminal 6).

Figure 7–37. Shunt regulator (Harris Semiconductor, *Linear & Telecom ICs,* 1991, p. 2–12)

7.8 Positive Voltage Regulators

Figure 7–38 shows the pinouts and functional block diagram for typical positive voltage regulators (the Harris Semiconductor CA3085/CA3085A/CA3085B). Table 7–4 shows the key design characteristics. Notice that the internal voltage reference (typical 1.6 V) is permanently connected to the noninverting input of the frequency-compensated error amplifier. Such regulators provide much better regulation than the three-terminal or hybrid devices discussed thus far in this chapter. As shown in Table 7–4, typical line and load regulation is about 0.025%.

These regulators provide output voltages ranging from 1.7 to 46 V at currents up to 100 mA. As shown in Fig. 7–38, the regulators contain features such as frequency compensation, short-circuit protection, temperature-compensated reference voltage, current limiting, and booster input. Quiescent current is low (typically 3 to 4 mA), as is temperature coefficient for both V_{REF} and the output voltage (typically 0.0035%/°C). Ripple rejection is typically 50 dB, and can be increased when an external capacitor is used.

These regulators are useful in a wide range of applications for regulating high-current, switching, shunt, and positive or negative voltages. The devices are also applicable for current and dual-tracking regulation. The CA3085A and B have output-current capabilities up to 100 mA, while the CA3085 provides up to 12 mA without an external pass transistor. All three devices can provide voltage regulation at load currents greater than 100 mA with a suitable external pass transistor. The CA3085 series has an unregulated input voltage range from 7.5 to 30 V. The

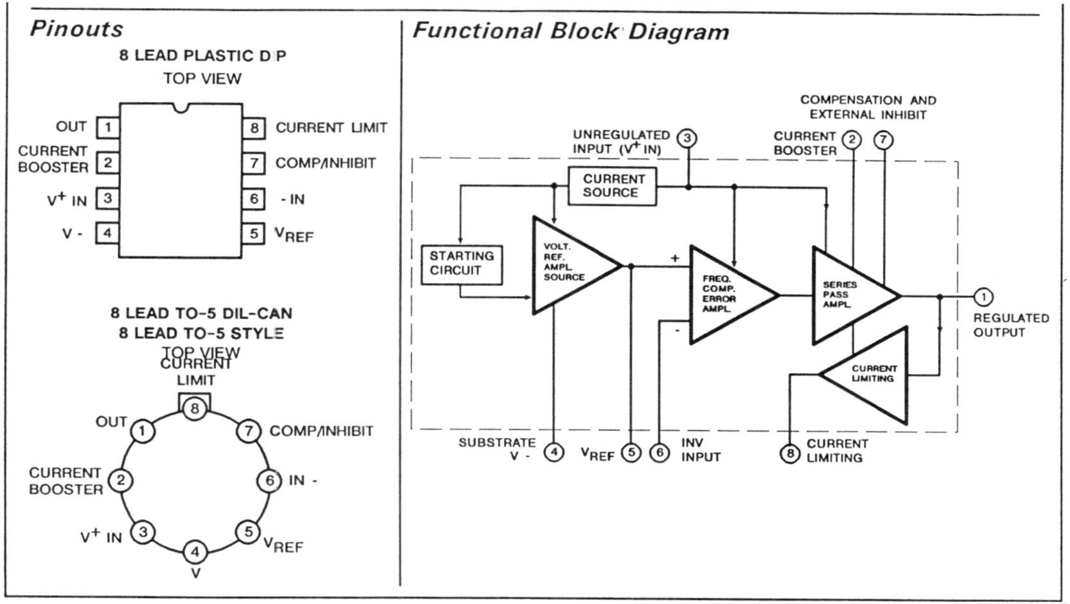

CAUTION: These devices are sensitive to electrostatic discharge. Proper I.C. handling procedures should be followed.

Figure 7–38. Pinouts and functional diagram for CA3085/A/B positive voltage regulator (Harris Semiconductor, *Linear & Telecom ICs,* 1991, p. 2–39)

Table 7-4. Key design characteristics for CA3085/A/B regulators (Harris Semiconductor, *Linear & Telecom ICs*, 1991, p. 2–41)

ELECTRICAL CHARACTERISTICS

Characteristics	Symbol	Test Conditions: Test Circuit Fig. No.	Test Conditions: $T_A = 25°C$ (Unless indicated otherwise)	CA3085 MIN.	CA3085 TYP.	CA3085 MAX.	CA3085A MIN.	CA3085A TYP.	CA3085A MAX.	CA3085B MIN.	CA3085B TYP.	CA3085B MAX.	UNITS
Reference Voltage	V_{REF}	4	$V^+_{IN} = 15V$	14	16	18	15	16	17	15	16	17	V
Quiescent Regulator Current	$I_{quiescent}$	4	$V^+_{IN} = 30V$	—	3.3	4.5	—	—	—	—	—	—	mA
			$V^+_{IN} = 40V$	—	—	—	—	3.65	5	—	—	—	
			$V^+_{IN} = 50V$	—	—	—	—	—	—	—	4.05	7	
Input Voltage Range	V_{IN}(range)	—	—	7.5	—	30	7.5	—	40	7.5	—	50	V
Maximum Output Voltage	V_O(max.)	4	$V^+_{IN} = 30, 40, 50$ V# $R_L = 365\Omega$ Term. No. 6 to Gnd	26	27	—	36	37	—	46	47	—	V
Minimum Output Voltage	V_O(min.)	4	$V^+_{IN} = 30V$	—	1.6	1.8	—	1.6	1.7	—	1.6	1.7	V
Input Output Voltage Differential	$V_{IN}V_{OUT}$	—	—	4	—	28	4	—	38	3.5	—	48	V
Limiting Current	I_{LIM}	7	$V^+_{IN} = 16V$ $V^+_{OUT} = 10V$ $R_{SCP}• = 6\Omega$	—	96	120	—	96	120	—	96	120	mA
Load Regulation▲	—	—	$I_L = 1$ to 100 mA. $R_{SCP} = 0$	—	—	—	—	0.025	0.15	—	0.025	0.15	%V_{OUT}
		—	$I_L = 1$ to 100 mA. $R_{SCP} = 0$ $T_A = 0°C$ to 70°C	—	—	—	—	0.035	0.6	—	0.035	0.6	
		—	$I_L = 1$ to 12 mA. $R_{SCP} = 0$	—	0.003	0.1	—	—	—	—	—	—	

Characteristics	Symbol	Test Conditions: Test Circuit Fig. No.	Test Conditions: $T_A = 25°C$ (Unless indicated otherwise)		Limits: CA3085 MIN.	CA3085 TYP.	CA3085 MAX.	CA3085A MIN.	CA3085A TYP.	CA3085A MAX.	CA3085B MIN.	CA3085B TYP.	CA3085B MAX.	UNITS
Line Regulation•	—	—	$I_L = 1$ mA. $R_{SCP} = 0$		—	0.025	0.1	—	0.025	0.075	—	0.025	0.04	%/V
		—	$I_L = 1$ mA. $R_{SCP} = 0$, $T_A = 0°C$ to 70°C		—	0.04	0.15	—	0.04	0.1	—	0.04	0.08	
Equivalent Noise Output Voltage	V_{NOISE}	11	$V^+_{IN} = 25V$	$C_{REF} = 0$	—	0.5	—	—	0.5	—	—	0.5	—	mV p p
				$C_{REF} = 0.22\mu F$	—	0.3	—	—	0.3	—	—	0.3	—	
Ripple Rejection	—	12	$V^+_{IN} = 25V$, f = 1kHz	$C_{REF} = 0$	—	50	—	—	50	—	45	50	—	dB
				$C_{REF} = 2\mu F$	—	56	—	—	56	—	50	56	—	
Output Resistance	r_O	12	$V^+_{IN} = 25V$. f = 1kHz		—	0.075	11	—	0.075	0.3	—	0.075	0.3	Ω
Temperature Coefficient of Reference and Output Voltages	ΔV_{REF}, ΔV_O	—	$I_L = 0$. $V_{REF} = 1.6V$		—	0.0035	—	—	0.0035	—	—	0.0035	—	%/°C
Load Transient Recovery Time: Turn On	t_{ON}	16	$V^+_{IN} = 25V$. +50mA Step		—	1	—	—	1	—	—	1	—	μs
Turn Off	t_{OFF}		$V^+_{IN} = 25V$. −50mA Step		—	3	—	—	3	—	—	3	—	μs
Line Transient Recovery Time: Turn On	t_{ON}	—	$V^+_{IN} = 25V$. f = 1kHz. 2V Step		—	0.8	—	—	0.8	—	—	0.8	—	μs
Turn Off	t_{OFF}	—			—	0.4	—	—	0.4	—	—	0.4	—	μs

30V (CA3085), 40V (CA3085A), 50V (CA3085B)

° R_{SCP} Short circuit protection resistance

• Load Regulation = $\frac{\Delta V_{OUT}}{V_{OUT}(\text{initial})} \times 100\%$

▲Line Regulation = $\frac{(\Delta V_{OUT})}{(V_{OUT\,(\text{initial})})\,(\Delta V_{IN})} \times 100\%$

CA3085A and CA3085B have unregulated input voltages from 7.5 to 40 V and 50 V, respectively. The minimum regulated output voltages are 26 V (CA3085), 36 V (CA3085A) and 46 V (CA3085B). The following paragraphs describe some design examples for the line.

7.8.1 High-Current Voltage Regulator

Figure 7–39 shows the CA3085 connected in the basic voltage-regulator configuration. The output voltage depends on the ratio $(R_1 + R_2)/R_1$ times 1.6 V. As a first trial value, select R_1 and R_2 such that 1 mA is drawn in parallel with the load RL (10 k for a 10-V output, 5 k for a 5-V output, etc.). So for an output of 10 V, R_2 should drop the output by about 8.4 V, while R_1 drops the remaining 1.6 V to ground, and the inverting input is at 1.6 V. Use 1.6 k for R_1 and 8.4 k for R_2.

The value of the current-limiting resistor between pins 1 and 8 should be such that the voltage drop is about 0.5 to 0.7 V when the desired current-limit point is reached. As shown in Fig. 7–39, the value of the resistor is set at 5.6 ohms. Likewise, the value of the frequency-compensation capacitor between pins 6 and 7 is set at 100 pF.

If used, C_{REF} is connected between pin 5 and ground. As shown in Table 7–4, C_{REF} can be omitted if a noise voltage of 0.5 mV p-p and a ripple rejection of 50 dB can be tolerated. If a value of 2 μF is used for C_{REF}, the ripple rejection is increased to 56 dB. If 0.22 μF is used for C_{REF}, noise drops to 0.3 mV p-p.

For simplified design, consider the total device dissipation to be the product of the input-output voltage differential and output current. Ignore the other factors such

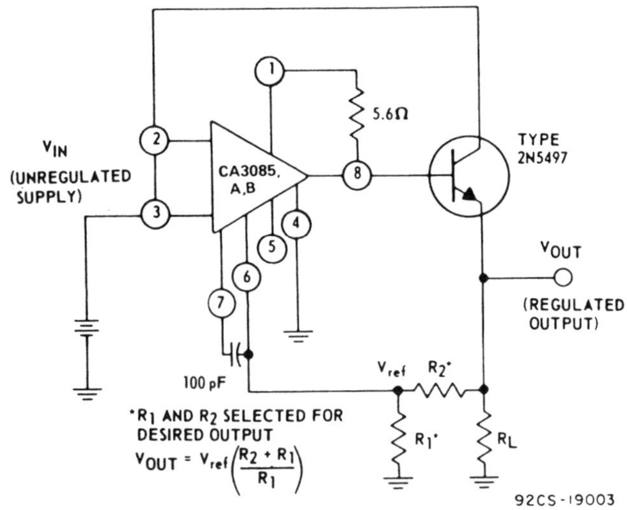

Figure 7–39. High-current voltage regulator (Harris Semiconductor, *Linear & Telecom ICs,* 1991, p. 2–45)

as quiescent current. According to the data-sheet graphs for the CA3085, the device can dissipate 630 mW at temperatures up to 55°C (without a heat sink) and 1.6 W (with a heat sink, but only if the TO-5 package is used, since there is no suitable heat sink for a DIP package). Above an ambient temperature of 55°C, derate linearly at about 7 mW/°C (with no heat sink) or 16.7 mW/°C (with heat sink).

If the full 100-mA output capability of the CA3085A or B is required, the data-sheet graphs show that the maximum input-output differential is about 5 V. The maximum output drops off to about 40 mA for a differential of 13 V, and to 20 mA for a differential of 23 V. These figures assume no heat sink for the CA3085A or B. If a heat sink is used, then the full 100 mA is available with a differential of about 15 V, and 85 mA is possible with a differential of 18 V or less. So always design for a minimum input-output differential, if practical.

Keep in mind that the maximum load current and input-output differential for the overall circuit of Fig. 7–39 depend on the characteristics of the booster transistor (a 2N5497 in this case), not on the CA3085. Therefore, it is essential that you check the data-sheet characteristics (maximum voltage, current, dissipation, derating, etc.) in addition to the IC regulator characteristics.

7.8.2 Current Regulator

Figure 7–40 shows the CA3085 used as a current regulator. The design considerations described thus far apply, except for resistor R_1. As shown by the equation, the value of R_1 depends on the desired current capability and the internal reference voltage (1.6 V). For example, for a load that produces the full 2 A of current, the value of R_1 is 1.6/2 = 0.8 ohm. Use at least a 5-W resistor for R_1, since the full current must pass through R_1 (I^2R, or 4 × 0.8 = 3.2 W). Again, remember that the maximum load, voltage, heat dissipation, etc., depend on the characteristics of transistor Q_1, so start by checking out the Q_1 data sheet.

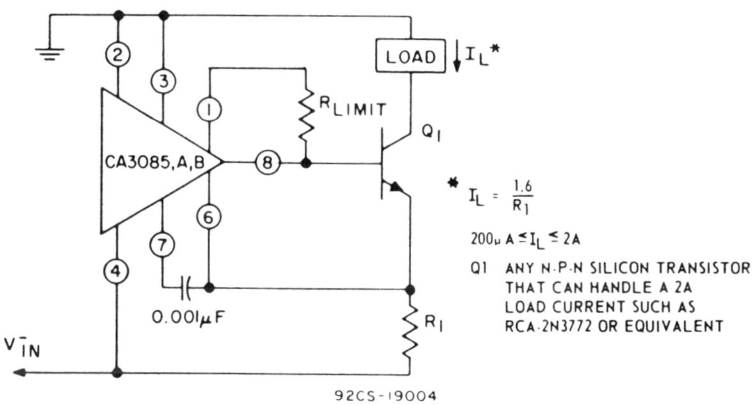

Figure 7–40. Current regulator (Harris Semiconductor, *Linear & Telecom ICs,* 1991, p. 2–45)

7.8.3 Combination Positive/Negative Regulator

Figure 7–41 shows two CA3085s connected to provide regulation for both positive and negative voltages. Since the CA3085 is a positive regulator, the position portion of the circuit is standard, and all of the design considerations discussed thus far apply. Use 5 to 7 ohms for R_{SCP} to produce a 0.5- to 0.7-V drop when about 100 mA is reached. Select R_1 and R_2 so that 1 mA is drawn in parallel with the load (10 k for 10-V output, with R_1 at 1.6 k and R_2 at 8.4 k).

In the negative supply, use 5 to 7 ohms for R_{SCP} to produce a 0.5- to 0.7-V drop across the base and emitter of Q_1 when about 100 mA is reached. Again, select R_1 and R_2 so that 1 mA is drawn in parallel with the load, but use the equation

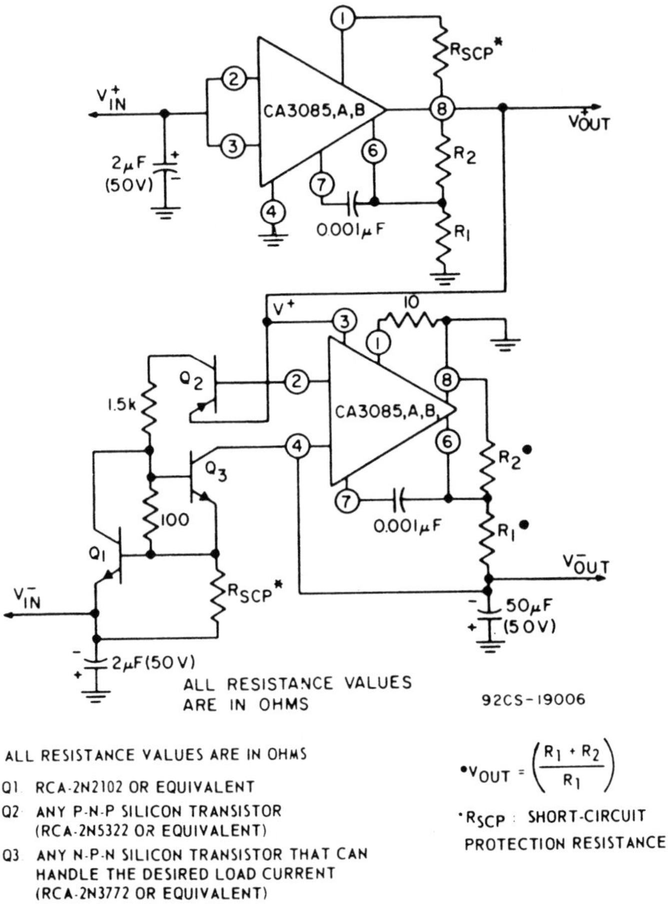

Figure 7–41. Combination positive/negative regulator (Harris Semiconductor, *Linear & Telecom ICs*, 1991, p. 2–45)

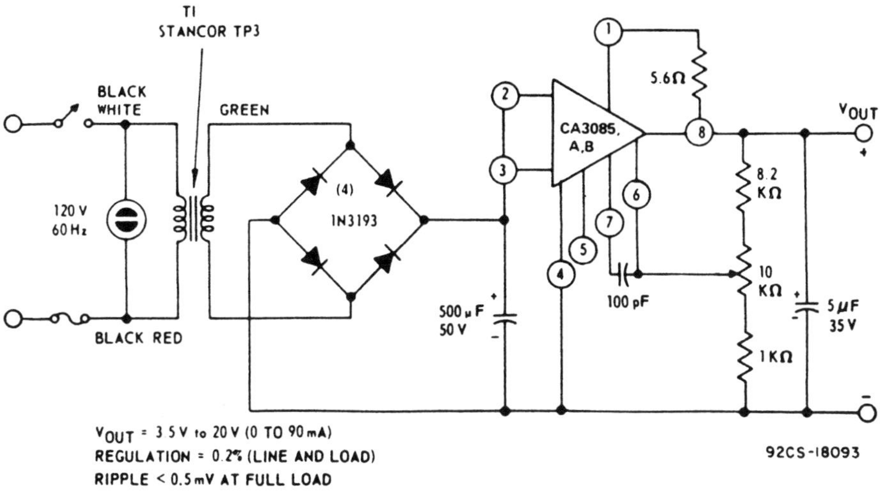

Figure 7–42. Complete power supply (Harris Semiconductor, *Linear & Telecom ICs,* 1991, p. 2–41)

shown to calculate the values. For example, if the output is to be –15 V, use a total of 15 k for $R_1 + R_2$. As a first trial value let R_1 equal 1 k, and R_2 equal the desired output voltage less R_1, or 14 k in this case.

7.8.4 Complete Power Supply

Figure 7–42 shows the CA3085 connected as the regulator for a complete power supply. Since all of the circuit values have been provided, no design calculations are required. However, it is always practical to check the dissipation characteristics for possible problems. For example, since the circuit is capable of 20 V regulated output, the unregulated input at pin 3 must be 20 V or greater. When the circuit is adjusted for 3.5 V output, there is a differential of about 16 or 17 V across the CA3085. As discussed in Section 7.8.1, the maximum output current drops off to about 40 mA for a differential of 13 V, without a heat sink. With a heat sink, 100 mA is available with a 15-V differential, and 85 mA is available with an 18-V differential. So if a heat sink is used, the rated 90 mA is realistic for the circuit.

7.9 Programmable Micropower Regulator

Figure 7–43 shows the functional block diagram and pin configuration for a CMOS programmable micropower positive voltage regulator (the Harris Semiconductor ICL7663S). Table 7–5 shows the key design characteristics. The device is a lower-power, high-efficiency, positive voltage regulator that accepts 1.6- to 16-V inputs and produces adjustable outputs from 1.3 to 16 V at currents up to 40 mA.

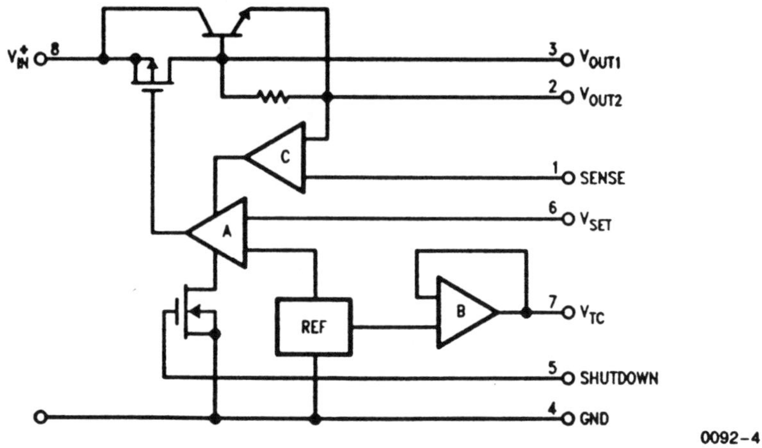

Figure 7–43. ICL7663S CMOS programmable micropower regulator (Harris Semiconductor, *Linear & Telecom ICs,* 1991, p. 2–126)

The ICL7663S is well suited for battery-powered supplies, featuring 4 μA quiescent current, low input-output differential, output current-sensing, and logic-level shutdown control. In addition, the device has a negative temperature-coefficient output suitable for generating a temperature-compensated display-drive voltage for LCD displays. Because of the unique operating characteristics of this device, we discuss some important design considerations before getting into design examples.

7.9.1 Functional Description

The ICL7663S is a CMOS IC incorporating all the functions of a voltage regulator plus protection circuits on a single chip. As shown in Fig. 7–43, the main blocks are a bandgap-type voltage reference, an error amplifier, and an output driver with both PMOS and NPN pass transistors.

The reference voltage, trimmed to 1.29 V ± 15 mV for the ICL7663SA, and the input voltage at the V_{SET} terminal (from the divider network) are compared in amplifier A. In turn, error amplifier A drives a P-channel pass transistor that is sufficient for low currents (under about 5 mA). When currents exceed 5 mA, the current is passed by the NPN transistor connected as a follower. This configuration gives more gain and lower output impedance.

Logic-controlled shutdown is implemented with an N-channel transistor. Current sensing is achieved by comparator C, which monitors both the SENSE and V_{OUT2} terminals. The device has an output (V_{TC}) from buffer amplifier B. The V_{TC} output can be used with amplifier A to generate programmable-temperature-coefficient output voltages.

The amplifier, reference, and comparator circuits all operate at bias level well below 1 μA, resulting in very low quiescent current (typically 4 μA, with a maxi-

Table 7–5. Key design characteristics for ICL7663S regulator (Harris Semiconductor, *Linear & Telecom ICs,* 1991, p. 2–125)

ELECTRICAL CHARACTERISTICS Specifications below applicable to both ICL7663S and ICL7663SA unless otherwise stated. V^+_{IN} = 9V, V_{OUT} = 5V, T_A = 25°C, unless otherwise stated. See Test Circuit, Figure 3.

Symbol	Parameter	Test Conditions	Limits			Units
			Min	Typ	Max	
V^+_{IN}	***Input Voltage***	***ICL7663S*** T_A = 25°C 0°C < T_A < +70°C −25°C < T_A < +85°C ***ICL7663SA*** 0°C < T_A < +70°C −25°C < T_A < +85°C	 1.5 1.6 1.6 ***1.6*** ***1.6***		 ***16*** ***16*** ***16*** ***16*** ***16***	 V V V
I_Q	***Quiescent Current***	1.4V ≤ V_{OUT} ≤ 8.5V, No Load V^+_{IN} = 9V, 0°C < T_A < 70°C −25°C < T_A < +85°C V^+_{IN} = 16V, 0°C < T_A < +70°C −25°C < T_A < +85°C			 ***10*** ***10*** ***12*** ***12***	 µA µA
V_{SET}	***Reference Voltage***	I_{OUT1} = 100 µA, V_{OUT} = V_{SET} ***ICL7663S*** T_A = 25°C ***ICL7663SA*** T_A = 25°C	 1.2 ***1.275***	 1.3 ***1.29***	 1.4 ***1.305***	 V V
$\frac{\Delta V_{SET}}{\Delta T}$	***Temperature Coefficient***	0°C < T_A < +70°C −25°C < T_A < +85°C		***100*** ***100***		ppm ppm
$\frac{\Delta V_{SET}}{V_{SET} \cdot \Delta V_{IN}}$	***Line Regulation***	2V < V_{IN} < 15V 0°C < T_A < +70°C −25°C < T_A < +85°C		 0.03 0.03	 ***0.3***	 %/V %/V
I_{SET}	***V_{SET} Input Current***	0°C < T_A < +70°C −25°C < T_A < +85°C		0.01 0.01	***10*** ***10***	nA nA
I_{SHDN}	Shutdown Input Current			±0.01	10	nA
V_{SHDN}	Shutdown Input Voltage	V_{SHDN} HI: Both V_{OUT} Disabled V_{SHDN} LO: Both V_{OUT} Enable	1.4		 0.3	V V
I_{SENSE}	Sense Pin Input Current			0.01	10	nA
V_{CL}	Sense Pin Input Threshold			0.5		V
R_{SAT}	***Input-Output Saturation Resistance (Note 3)***	V^+_{IN} = 2V, I_{OUT1} = 1 mA V^+_{IN} = 9V, I_{OUT1} = 2 mA V^+_{IN} = 15V, I_{OUT1} = 5 mA		***170*** ***50*** ***35***	***350*** ***100*** ***70***	Ω Ω Ω
$\frac{\Delta V_{OUT}}{\Delta I_{OUT}}$	***Load Regulation***	1 mA < I_{OUT2} < 20 mA 50 µA < I_{OUT1} < 5 mA		1 2	***3*** ***10***	Ω Ω
I_{OUT2}	***Available Output Current (V_{OUT2})***	3V ≤ V_{IN} ≤ 16V, V_{IN} − V_{OUT2} = 1.5V	***40***			mA
V_{TC}	Negative Tempco Output (Note 4)	Open-Circuit Voltage		0.9		V
I_{TC}		Maximum Sink Current	0	8	2.0	mA
$\frac{\Delta V_{TC}}{\Delta T}$	Temperature Coefficient	Open Circuit		+2.5		mV/°C
$I_{L(MIN)}$	Minimum Load Current	(Includes V_{SET} Divider) T_A = 25°C 0°C < T_A < +70°C −25°C < T_A < +85°C		 0.2 0.2	 ***1.0*** ***5.0*** ***5.0***	 µA µA µA

NOTE: 1. Connecting any terminal to voltages greater than (V^+_{IN} + 0.3V) or less than (GND − 0.3V) may cause destructive device latch-up. It is recommended that no inputs from sources operating on external power supplies be applied prior to ICL7663S power-up.

2. Derate linearly above 50°C at 5 mW/°C for Plastic Minidip, 7.5 mW/°C for TO-99 can, and 10 mW/°C for CERDIP.

3. This parameter refers to the saturation resistance of the MOS pass transistor. The minimum input-output voltage differential at low current (under 5 mA), can be determined by multiplying the load current (including set resistor current, but not quiescent current) by this resistance.

4. This output has a positive temperature coefficient. Using it in combination with the inverting input of the regulator at V_{SET}, a negative coefficient results in the output voltage. See Figure 5 for details. Pin will not source current.

5. All pins are designed to withstand electrostatic discharge (ESD) levels in excess of 2000V.

6. All significant improvements over the industry standard ICL7663 are highlighted in ***bold italics***.

mum of 12 μA). This low current does limit the dynamic response of the circuits, however, and transients are best corrected outside the regulator feedback loop.

7.9.2 Basic Precautions

The ICL7663S is designed to regulate battery voltages in the 5- to 15-V region, at maximum load currents of about 5 to 30 mA (although the device is capable of 40 mA). Although intended for low-power operation, power-dissipation limits must be observed. For example, the power dissipation in case of a 10-V supply that is regulated down to 2 V, with a load current of 30 mA (or 240 mW), exceeds the power-dissipation rate of both the SOIC and Minidip package configurations (200 mW). However, the device is available in a CERDIP package with a rated power dissipation of 500 mW.

Since the device is CMOS, all of the usual CMOS precautions must be observed. For example, every input pin must go somewhere. Equally important, the maximum values of applied voltages and current limits must be carefully observed. (See Note 1 in Table 7–5.) Also notice that the test circuit, called Figure 3 in Table 7–5, is shown in Fig. 7–44. Failure to observe the limits and notes given in Table 7–5 can cause circuit failure and possible component damage.

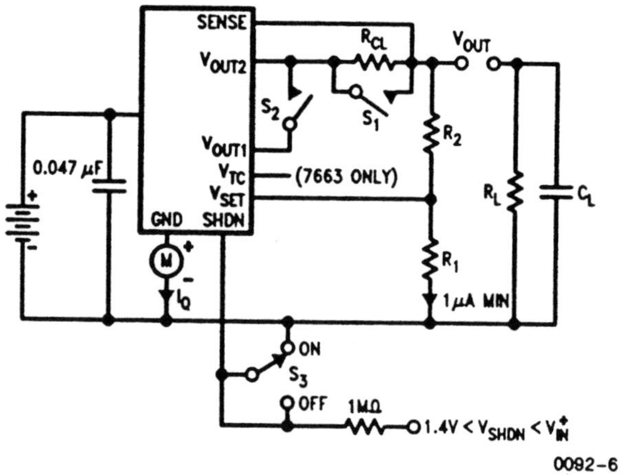

NOTE: 1. S_1 when closed disables output current limiting.

2: Close S_2 for V_{OUT1}, open S_2 for V_{OUT2}.

3: $$V_{OUT} = \frac{R_2 + R_1}{R_1} V_{SET}.$$

4: I_Q quiescent current is measured at GND pin by meter M.

5: S_3 when ON, permits normal operation, when OFF, shuts down both V_{OUT1} and V_{OUT2}.

Figure 7–44. ICL7663S test circuit (Harris Semiconductor, *Linear & Telecom ICs,* 1991, p. 2–128)

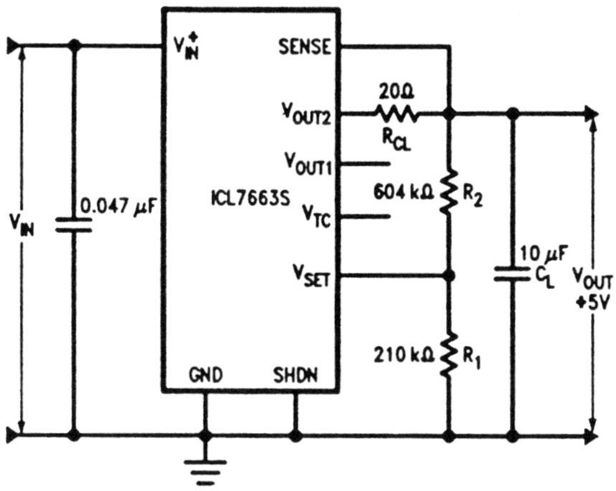

0092-7

$$V_{OUT} = \frac{R_2 + R_1}{R_1} V_{SET} = 5V$$

$$I_{CL} = \frac{V_{CL}}{R_{CL}} = 25 \text{ mA}$$

Figure 7–45. Positive regulator with current limit (Harris Semiconductor, *Linear & Telecom ICs,* 1991, p. 2–128)

Figure 7–45 shows the ICL7663S in the basic regulator configuration (a positive regulator with current limit). The following sections describe the design considerations for the basic circuit.

7.9.3 Input Voltages

The device accepts working inputs of 1.5 V to 16 V. When power is applied, the rate-of-rise of the input may be hundreds of volts per microsecond. This is potentially harmful to the regulator, where internal operating currents are in the nanoampere range. The 0.047-µF capacitor connected across the input will limit inputs to a safe level of about 2 V/µs. Use the capacitor in all applications. In severe rate-of-rise cases, it may be advisable to use an RC network on the shutdown terminal (pin 5) to delay output turn-on. Battery-charging surges, transients, and assorted noise signals should be kept from the regulator by RC filtering, zener protection, or even fusing (in extreme cases).

7.9.4 Output Voltages

The resistor divider R_2/R_1 is used to scale the reference voltage V_{SET} using the equation shown in Fig. 7–45. If more convenient, use the alternate equation $V_{OUT} = (1 + R_2/R_1)\, V_{SET}$. The ICL7663SA has a guaranteed V_{SET} of 1.29 V ± 15 mA. If 1% resistors are used for R_1 and R_2, the initial output voltage (as calculated by either equation) will be within ±2.7% of the desired value. However, if greater precision is required, use a potentiometer for R_1 and R_2, adjust for the desired output, then disconnect the pot and measure the resistances. Substitute fixed resistances of the same value, using at least 1% resistors.

The low leakage current of the V_{SET} terminal (pin 6) allows R_1 and R_2 to be tens of megohms for minimum quiescent current. However, some load current is required for proper operation, so for extremely low-drain applications, it is necessary to draw at least 1 μA. This can include the current for R_1 and R_2.

The low limit of output voltage is set by V_{SET} (1.29 V), while the high limit is set by the input-output differential. Output voltages up to nearly the V_{IN} supply may be obtained (at low load currents). Use the V_{OUT1} terminal (pin 3) for minimum input-output differential. The input-output differential increases to 1.5 V when using V_{OUT2} (pin 2).

7.9.5 Output Currents

Output currents of less than 5 mA are obtained with the least input-output differential at the V_{OUT1} terminal. (It is recommended that V_{OUT1} be connected to V_{OUT2} in this case.) Outputs greater than 5 mA should be taken from V_{OUT2} (with V_{OUT1} left open). Notice that the absolute maximum output from V_{OUT1} is 30 mA, while the maximum from V_{OUT2} is 50 mA. These high currents can be obtained only as far as the package dissipation allows (refer to Section 7.9.2). It is strongly recommended that output current-limit sensing be used when output currents near the maximum are involved.

7.9.6 Output Current-Limit Sensing

The on-chip comparator (C in Fig. 7–43) permits shutdown of the regulator output in the event of excessive current drain. As shown in Fig. 7–45, a current-limiting resistor R_{CL} is placed in series with V_{OUT2}, and the SENSE terminal is connected to the load side of R_{CL}. When the current through R_{CL} is high enough to produce a voltage drop equal to V_{CL} (about 0.5 V), the voltage feedback is bypassed and the regulator output is limited to this current. Simply divide V_{CL} by the desired load current (I_{CL}) to find R_{CL}.

7.9.7 Logic-Level Shutdown

The ICL7663S can be shut down by a logic-level control signal applied at pin 5. Since pin 5 is an input (to a CMOS IC), the pin should be tied to ground if not used. A voltage level of less than 0.3 V at pin 5 will keep the regulator on, while a voltage greater than 1.4 V turns the regulator off. When off, the quiescent current is typically less than 4 μA.

The pin-5 voltage should always be less than V_{IN}. If there is a possibility that the control signal could exceed V_{IN} (say, where the control signal originates from a system powered separately from the regulator), connect a 1-M resistor in series between the control-signal source and pin 5. This should limit the control-signal current to less than 100 μA.

7.9.8 Load Capacitor

The regulator has poor rejection of voltage fluctuations from sources about 10 Hz and above. Where this might be a problem, use a load capacitor C_L as shown in Fig. 7–45. The 10-μF value is suitable for most applications where the output is in the 5-V range. A lower-value load capacitor can be used if the output voltage is increased, but this is generally of little practical value.

7.9.9 Negative-TC Output Voltage

The ICL7663S has an additional output (V_{TC} at pin 7) that is 0.9 V relative to ground. This V_{TC} output has a temperature coefficient of +2.5 mV/°C. By applying the V_{TC} output to the inverting input of amplifier A (V_{SET} pin 6, Fig. 7–43), it is possible to produce output voltages with a negative temperature coefficient. Figure 7–46 shows the equivalent circuit and equations involved. Keep in mind that V_{REF} is an internal connection, R_3 is connected between pins 6 and 7, R_1 is connected between pin 6 and ground, while R_2 is connected between pin 6 and the selected output (pin 2 or 3, or both).

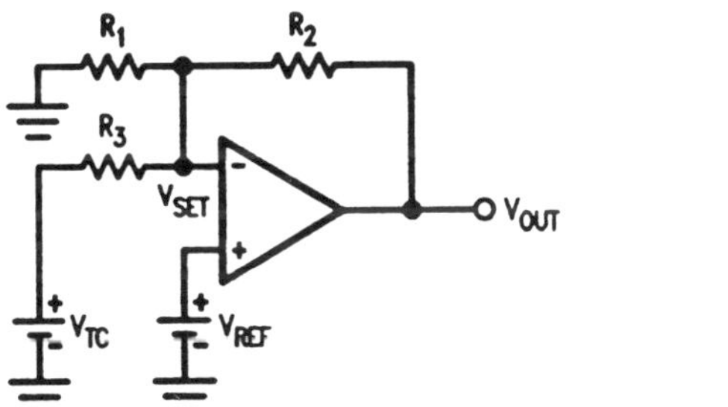

EQ. 1: $V_{OUT} = V_{SET}\left(1 + \frac{R_2}{R_1}\right) + \frac{R_2}{R_3}(V_{SET} - V_{TC})$

EQ. 2: $TC\ V_{OUT} = -\frac{R_2}{R_3}(TC\ V_{TC})$ in mV/°C

Where: $V_{SET} = 1.3V$
$V_{TC} = 0.9V$
$TCV_{TC} = +2.5$ mV/°C

Figure 7–46. Generating negative-TC output voltages (Harris Semiconductor, *Linear & Telecom ICs,* 1991, p. 2–129)

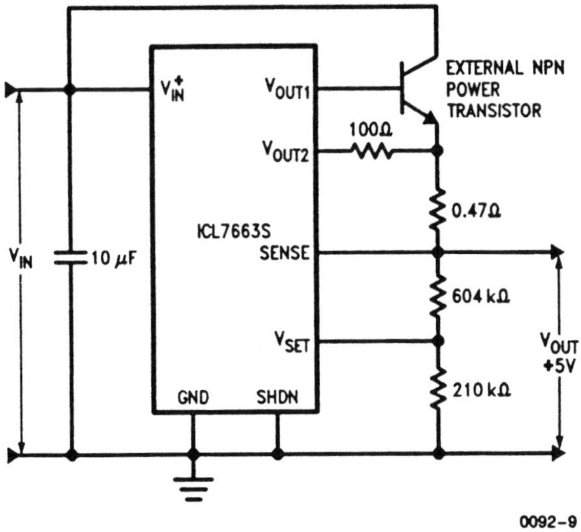

Figure 7–47. Boosting output current with external transistor (Harris Semiconductor, *Linear & Telecom ICs,* 1991, p. 2–129)

7.9.10 Boosting Output Current

Figure 7–47 shows the ICL7663S connected as the control element for an external NPN transistor. This circuit is essentially the same as shown in Fig. 7–45, except that the output current is limited only by the capacity of the external transistor. As discussed thus far in this chapter, consider the transistor dissipation as being the product of input-output differential and load current.

7.9.11 Temperature-Compensated Display-Drive Voltage

Figure 7–48 shows the ICL7663S connected to generate a temperature-compensated display voltage (V_{DSP}) for an LCD driver (the Harris Semiconductor ICM7233). Temperature has an important effect in the variation of threshold voltage in multiplexed LCD displays. As temperature rises, the threshold voltage goes down. For applications where the display temperature varies widely, a stable display voltage is essential.

7.10 Computer-Controlled Linear Regulators

It is possible to place a linear regulator under computer control with a relatively simple circuit. With such a circuit, the power supply output can be trimmed to a particular variation above (and below) the design-center voltage. If desired, the computer can run the trimmed output through a given sequence, stopping at selected output voltages. Any deviation from normal operation, at any particular output volt-

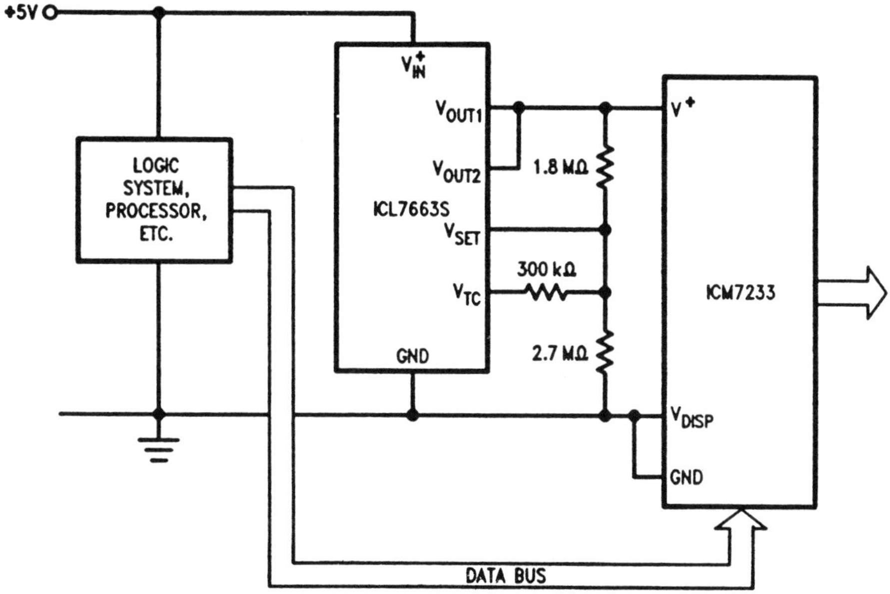

Figure 7–48. Generating TC display voltage (VDSP) for an LCD driver (Harris Semiconductor, *Linear & Telecom ICs,* 1991, p. 2–130)

age, will serve as a warning of impending malfunction or failure. This test approach can be used for diagnostics, for troubleshooting, and for engineering evaluation. The approach can also help detect skew, race conditions, timing problems, and noise or threshold problems. The following paragraphs describe computer-controlled regulators with both positive and negative outputs.

7.10.1 Computer-Controlled Positive Regulator

Figure 7–49 shows a standard three-terminal regulator (the National Semiconductor LM338) connected to provide a positive output under computer control. Table 7–6 shows the available trim range.

During normal operation, latch IC_1 is programmed so that Q_1/Q_2 outputs are high and Q_3/Q_4 outputs are low. Under these conditions, R_4 and R_5 are connected in parallel with R_6, and V_{OUT} is adjusted to 5 V with P_1. If Q_4 is commanded high by the computer, the net conductance from the adjust bus to ground decreases and V_{OUT} rises 3% to 5.15 V, as shown in Table 7–6. Conversely, if Q_1 is made low, V_{OUT} falls 3.3% to 4.835.

7.10.2 Computer-Controlled Negative Regulator

Figure 7–50 shows a standard three-terminal regulator (the National Semiconductor LM337) connected to provide a negative output under computer control. Table 7–7 shows the available trim range.

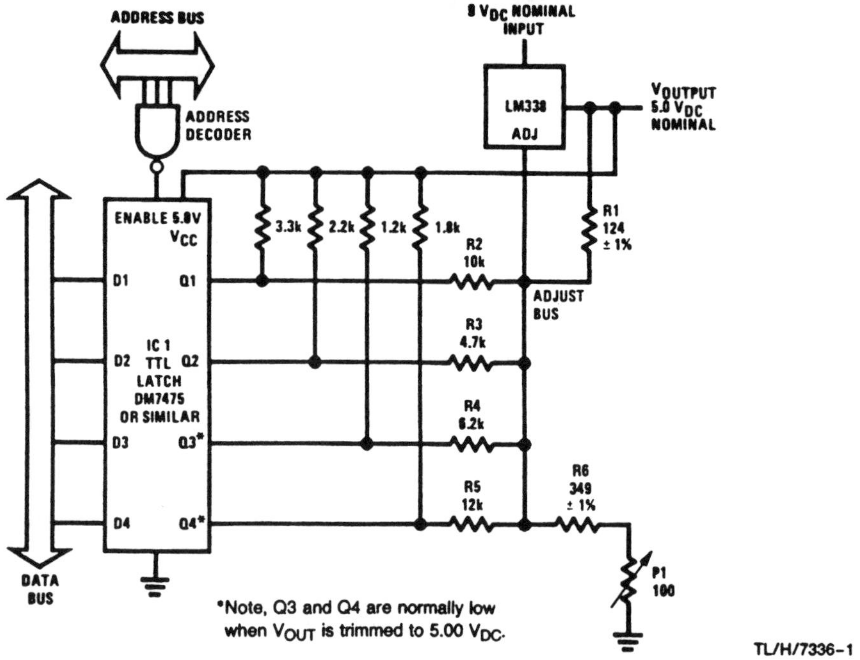

Figure 7–49. Computer-controlled positive regulator (National Semiconductor, *Linear Applications Handbook,* 1991, p. 1261)

Table 7–6. Available trim range for computer-controlled positive regulator (National Semiconductor, *Linear Applications Handbook,* 1991, p. 1261)

Q1	Q2	Q3	Q4	V_{OUT}	$\%\Delta V_{OUT}$
1	1	0	0	5.000V	(trimmed)
1	1	0	1	5.151V	+3.0%
1	1	1	0	5.299V	+6.0%
1	1	1	1	5.469V	+9.4%
0	1	0	0	4.835V	−3.3%
1	0	0	0	4.669V	−6.6%
0	0	0	0	4.526V	−9.5%

This circuit is well suited to ECL logic that uses –5.2 V. If the computer command is from TTL latches, the circuit will be suitable to interface between (0 V and 2.4 V) logic levels and the saturated PNP collectors. Resistors R_{101}–R_{104} are switched by transistors Q_{101}–Q_{104}. Notice that the resistors in Fig. 7–50 are in a binary-weighted proportion and provide a trim resolution in 3% steps. This makes it possible to set V_{OUT} within 2% of the desired value. To decrease V_{OUT} by 2%,

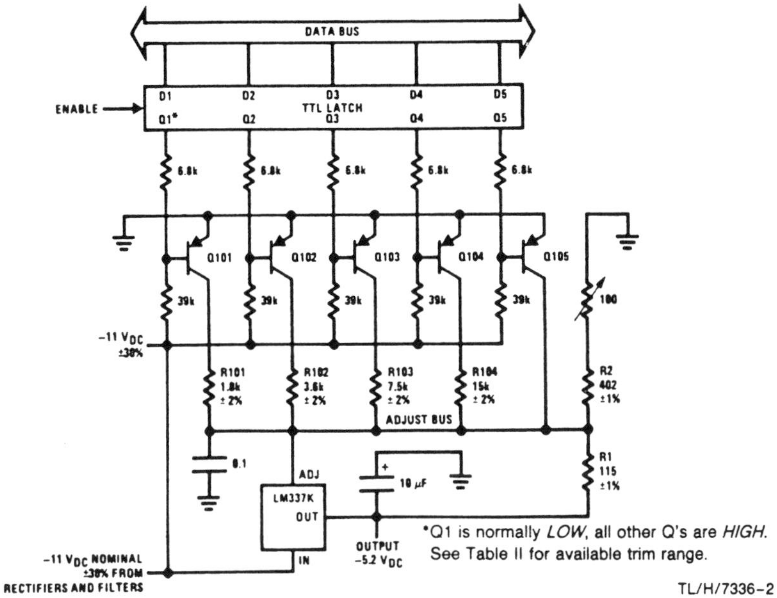

Figure 7–50. Computer-controlled negative regulator (National Semiconductor, *Linear Applications Handbook,* 1991, p. 1262)

Table 7–7. Available trim range for computer-controlled negative regulator (National Semiconductor, *Linear Applications Handbook,* 1991, p. 1262)

Q1	Q2	Q3	Q4	V_{OUT}	$\%\Delta V_{OUT}$
0	1	1	1	5.200V	(trimmed)
0	1	1	0	5.110V	−1.7%
0	1	0	1	5.025V	−3.4%
0	1	0	0	4.944V	−4.9%
0	0	1	1	4.853V	−6.7%
0	0	1	0	4.779V	−8.1%
0	0	0	1	4.707V	−9.5%
0	0	0	0	4.638V	−10.8%
1	0	0	0	5.310V	+2.1%
1	0	0	1	5.409V	+4.0%
1	0	1	0	5.513V	+6.0%
1	0	1	1	5.622V	+8.1%
1	1	0	0	5.757V	+10.7%
1	1	0	1	5.880V	+13.1%
1	1	1	0	6.010V	+15.6%
1	1	1	1	6.147V	+18.2%

change Q_4 to low. To increase V_{OUT} by 2%, set Q_1 high and Q_2, Q_3, Q_4 to low, in a standard offset-binary scheme.

The circuit of Fig. 7–50 provides another feature: controlled or automatic shutdown. If Q_{105} is made low, Q_{105} will saturate and pull the adjust bus to within 100 mV of ground. This causes V_{OUT} to collapse. In effect, the negative supply is shut down, and the computer draws substantially zero power.

7.11 Regulator with Ultra-Low Dropout

Low dropout voltage is a highly desirable feature for linear regulators used in battery-operated equipment (such as laptop computers). Usually, regulation must be maintained until the battery is almost entirely depleted. This section describes a simple regulator circuit with ultra-low dropout using a standard regulator (the Linear Technology LT1431). The circuit has only 85-mV dropout at 2.5 A. Dropout decreases drastically with lower output currents.

7.11.1 Basic Circuit Operation

Figure 7–51 shows the basic circuit connections. The LT1431 shunt regulator adjusts the output (COLLECTOR) to whatever value is required to force circuit out-

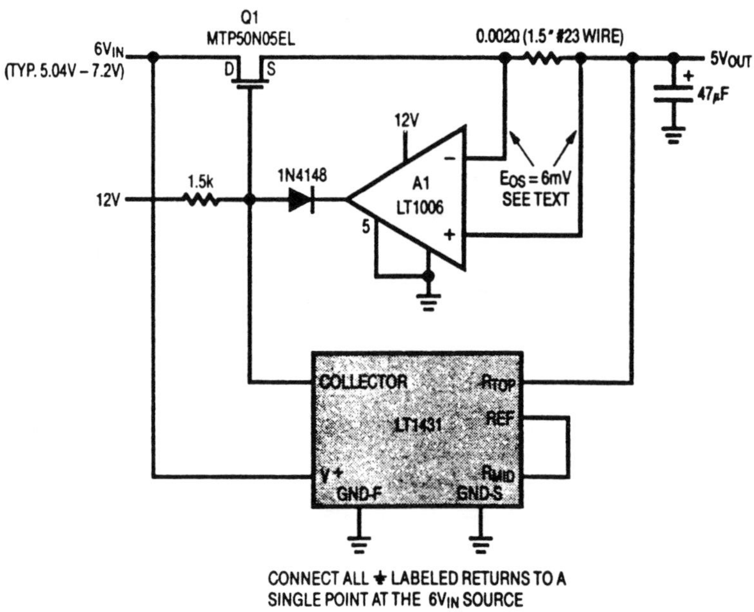

Figure 7–51. Ultra-low dropout regulator (Linear Technology, *Linear Applications Handbook,* 1990, p. DN32-1)

put to 5 V. The LT1431 internal trimming eliminates the usual feedback resistors and trimpots. The pass element Q_1 runs as a voltage-overdriven source follower. This configuration offers the lowest possible dropout voltage, although a +12-V bias source is required for the Q_1 gate. This +12-V source is commonly present in laptop computers and similar devices because of disk-drive and peripheral power requirements. Power drain on the +12-V supply is a few milliamps.

7.11.2 Dropout Characteristics

Figure 7–52 shows the dropout characteristics for the circuit with various types of Q_1. When the specified MTP50N05EL MOSFET is used for Q_1, dropout is 85 mA. This decreases to 8 mV at 0.25 A. Even with the higher-resistance Q_1 transistors shown, dropout is kept to very low levels.

7.11.3 Short-Circuit Protection

Figure 7–53 shows details of the low-resistance shunt used in Fig. 7–51. Table 7–8 shows resistance versus size for various wire gauges. The shunt is used by A_1 to sense current output. A 0.002-ohm shunt (1.5 in. of No. 23 wire) introduces 6 mV of drop when the circuit is at the maximum 3-A output limit.

A 6-mV current-limit trip point is derived by grounding the A_1 offset (pin 5). The 6-mV input-offset generated at A_1 is stable over time and temperature. Currents beyond 3 A cause the A_1 output to pull low, reducing Q_1 gate drive and shutting off the regulator output. Thus, under short-circuit overload conditions, A_1 and Q_1 form a linear current-control loop.

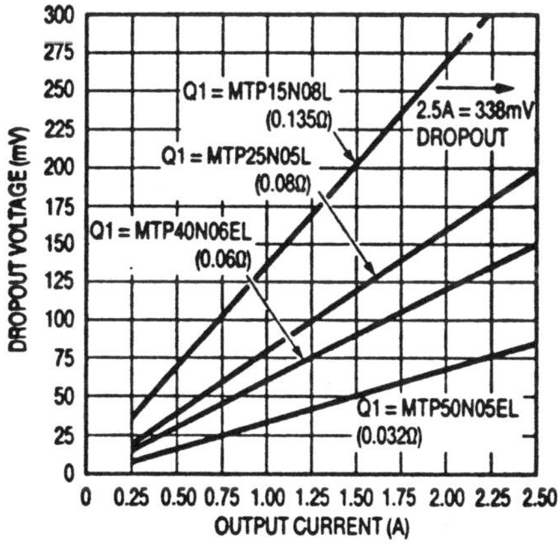

Figure 7–52. Dropout characteristics for ultra-low dropout regulator (Linear Technology, *Linear Applications Handbook,* 1990, p. DN32-2)

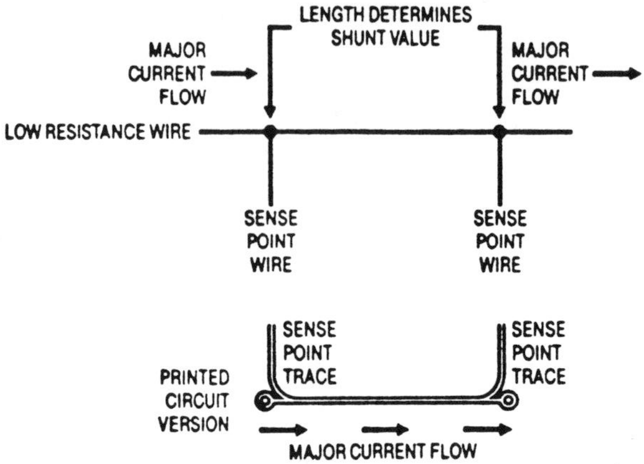

Figure 7–53. Details of low-resistance shunt for ultra-low dropout regulator (Linear Technology, *Linear Applications Handbook,* 1990, p. DN32-2)

Table 7–8. Resistance versus size for various copper-wire gauges (Linear Technology, Linear Applications Handbook, 1990, p. DN32–2)

Wire Gauge	*μΩ/Inch*
10	83
11	100
12	130
13	160
14	210
15	265
16	335
17	421
18	530
19	670
20	890
21	1000
22	1300
23	1700
24	2100
25	2700

7.12 Dual-Tracking Voltage Regulators

Figure 7–54 shows the pinout and functional block diagram for an adjustable dual-tracking regulator (the Raytheon RC4194). Table 7–9 shows the key design characteristics, both thermal and electrical. The RC4194 (and the Military Range RM1494) are dual-polarity tracking regulators designed to provide balanced or unbalanced positive and negative output voltages at currents up to 200 mA. A single

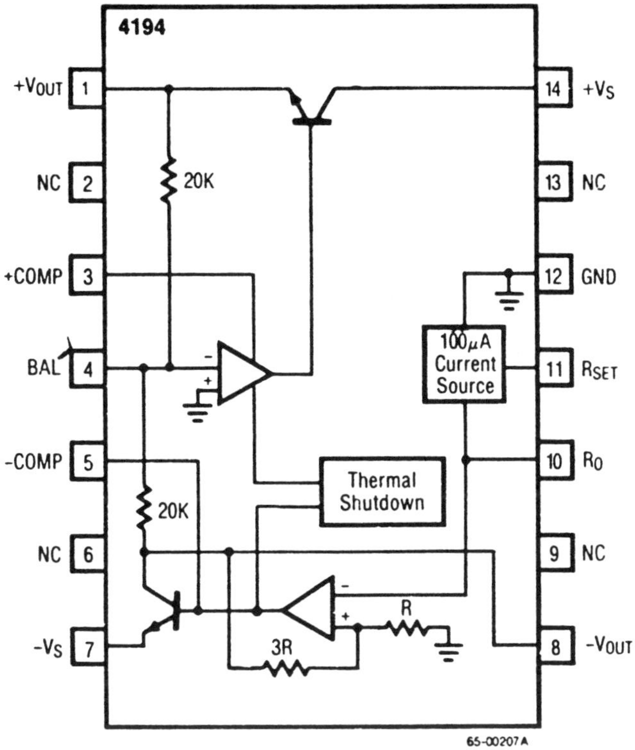

Figure 7–54. Pinout and functional diagram for RC4194 dual-tracking regulator (*Raytheon Linear Integrated Circuits,* 1989, p. 9–65)

external resistor adjustment can be used to change both outputs between the limits of ±50 mV and ±42 V.

These devices are designed for local on-card regulation, eliminating distribution problems associated with single-point regulation. To simplify application, the regulators require a minimum number of external parts. The regulators are available in three package types to accommodate various power requirements. The K (TO–66) power package can dissipate up to 3 W at an ambient temperature of +25°C. The D 14-pin DIP will dissipate up to 1 W, and the N 14-pin DOP dissipates up to 625 mW.

7.12.1 Basic Circuit Connections

Figure 7–55 shows the basic circuit connections, including compensation capacitors. The output voltage is set by the value of resistor R_S (at R_{SET} pin 11) and R_O at pin 10. The value of R_S sets the scale factor, while R_O sets the output voltage. As shown in Table 7–9, when R_S is set at 71.5 k, the scale factor is 2.5 k per volt. For example, for a 6-V output, make R_S = 71.5 k and R_O = 6 × 2.5 k = 15 k.

Table 7-9. Key design characteristics for RC4194 dual-tracking regulator (*Raytheon Linear Integrated Circuits,* 1989, p. 9–66)

Thermal Characteristics

	14-Lead Plastic DIP	14-Lead Ceramic DIP	9-Lead TO-66 Metal Can
Max. Junction Temp.	125°C	175°C	150°C
Max. P_D T_A <50°C	468mW	1042mW	2381mW
Therm. Res θ_{JC}	—	60°C/W	7°C/W
Therm. Res. θ_{JA}	160°C/W	120°C/W	42°C/W
For T_A >50°C Derate at	6.25 mW/°C	8.38 mW/°C	23.81 mW/°C

Electrical Characteristics ($\pm5 \le V_{OUT} \le V_{MAX}$; $-V_{IN} \le -8V$; $I_L = \pm1mA$; RM4194: $-55°C \le T_J \le +125°C$; RC4194: $0°C \le T_J \le +70°C$ unless otherwise specified)

Parameters	Test Conditions	Min	Typ	Max	Units
Line Regulation	$\Delta V_S = 0.1\ V_{IN}$		0.04	0.1	$\%V_{OUT}$
Load Regulation	4194K: $I_L < 200$ mA 4194D: $I_L < 100$ mA $\pm V_S = \pm(V_O + 5)V$		0.002	0.004	$\%V_O \times I_L$ (mA)
Output Voltage Drift With Temperature [3]					
Positive Output	$V_{OUT} = \pm5V$		0.002	0.015	%/°C
Negative Output	$V_{OUT} = \pm5V$		0.,003	0.015	%/°C
Supply Current[1] (Positive)	$V_S = \pm V_{MAX}$, $V_O = 0V$, $I_L = 0$ mA		+0.8	+2.5	mA
Supply Current[2] (Negative)	$V_S = \pm V_{MAX}$, $V_O = 0V$, IL = 0 mA		-1.8	-4.0	mA
Supply Voltage	RM4194	±9.5		±45	V
	RC4194	±9.5		±35	V
Output Voltage Scale Factor	$R_{SET} = 71.5\ k\Omega$, $T_J = +25°C$ $V_S = \pm V_{MAX}$	2.38	2.5	2.62	kΩ/V
Output Voltage Range	RM4194: $R_{SET} = 71.5\ k\Omega$, $I_L = 25$ mA	0.05		±42	V
	RC4194: $R_{SET} = 71.5\ k\Omega$, $I_L = 25$ mA	0.05		±42	V
Output Voltage Tracking			±0.4	±2.0	%
Ripple Rejection	F = 120 Hz, $T_J = +25°C$		70		dB
Input-Output Voltage Differential	$I_L = 50$ mA, $T_J = +25°C$	3.0			V
Short Circuit Current	$V_S = \pm30V$, $T_J = +25°C$		300		mA
Output Noise Voltage	$C_L = 4.7\ \mu F$, $V_O = \pm15V$ F = 10 Hz to 100 kHz		250		μV_{RMS}
Internal Thermal Shutdown			175		°C

Notes: 1. The current drain will increase by $50\mu A/V_{OUT}$ on positive side and $100\mu A/V_{OUT}$ on negative side.
2. The specifications above apply for the given junction temperatures since pulse test conditions are used.
3. Output voltage temperature drift guaranteed by design.

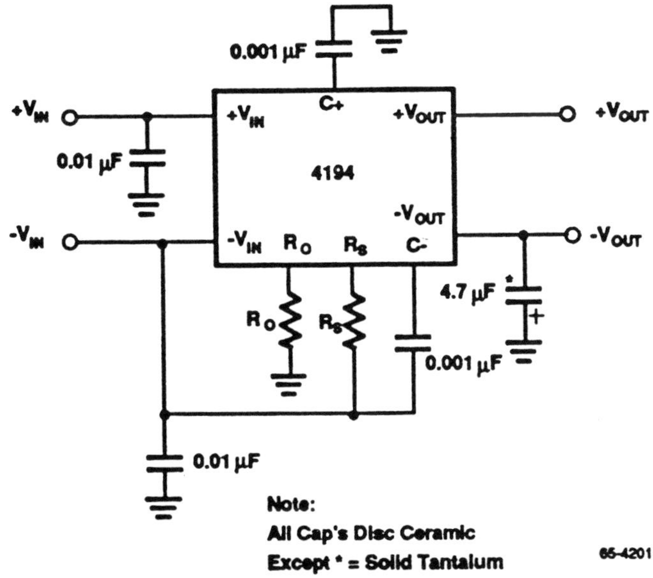

Figure 7–55. Basic circuit connections for dual-tracking regulator (*Raytheon Linear Integrated Circuits,* 1989, p. 9–71)

7.12.2 Compensation

For most applications, the compensation technique shown in the data sheet is sufficient. The positive section of the regulator is compensated by a 0.001-μF ceramic disc capacitor from the C+ terminal to ground. The negative regulator requires compensation at two points. The first is the C– terminal, which should have 0.001 μF to the $-V_{IN}$ pin ($-V_S$ or case for a TO–66 package). Use a ceramic disc here. The second compensation point for the negative section is the $-V_{OUT}$ pin. Here the capacitor should (ideally) be a 4.7-μF solid tantalum with enough reserve voltage capacity to avoid the momentary shorting and reforming that can occur with tantalum capacitors.

For systems where the cost of a solid tantalum capacitor cannot be justified, it is usually sufficient to use an aluminum capacitor with a 0.03-μF ceramic disc in parallel to bypass high frequencies. Also, if the rectifier filter capacitors have poor high-frequency characteristics (such as aluminum electrolytics) or if any impedance is in series with the $+V_{IN}$ and $-V_{IN}$ terminals, it is necessary to bypass these two points with 0.01-μF ceramic-disc capacitors.

As is the case with IC op-amps, some applications may not require these bypass capacitors. However, if in doubt, or if the trial circuit does not perform as desired, include all of the capacitors shown in Fig. 7–55. As always, all compensation and bypass capacitors should have short leads, solid grounds, and be located as close to the regulator as possible (at the IC pins if practical).

7.12.3 Diode Protection

Whenever any type of IC linear regulator is used, there are a number of conditions that, if not corrected, can destroy the regulator. For example, IC regulators can be destroyed by reversal of input- or output-voltage polarity, or if the input voltage drops below the output voltage in magnitude. Such conditions can be caused by inductive loads at the inputs or outputs of the regulator. Other problems are caused by heavy loads at the unregulated inputs to the regulator, which might cause the input voltage to drop below the output voltage at turnoff. If any of these problem conditions are present, protect the regulator using diodes.

Figure 7–56 shows six possible diode-protection points in the basic dual-tracking regulator circuit. These diodes should be high-speed types capable of handling large current surges. The diodes at the inputs and outputs prevent voltage at these points from becoming reversed. The diodes from outputs to inputs prevent the output voltage from exceeding the input voltage. In most cases, all six diodes are not needed. However, if in doubt, or if there are input-output problems during the trial-circuit stage of design, include all of the diodes shown in Fig. 7–56. Again, keep the diodes close to the regulator pins, with short leads.

7.12.4 Brownout Protection

Figure 7–57 shows a circuit for brownout protection for the ICs being supplied by the regulator. During brownout, line voltages may be reduced to as low as 75 V RMS, causing the input voltage to the regulator to drop below the minimum

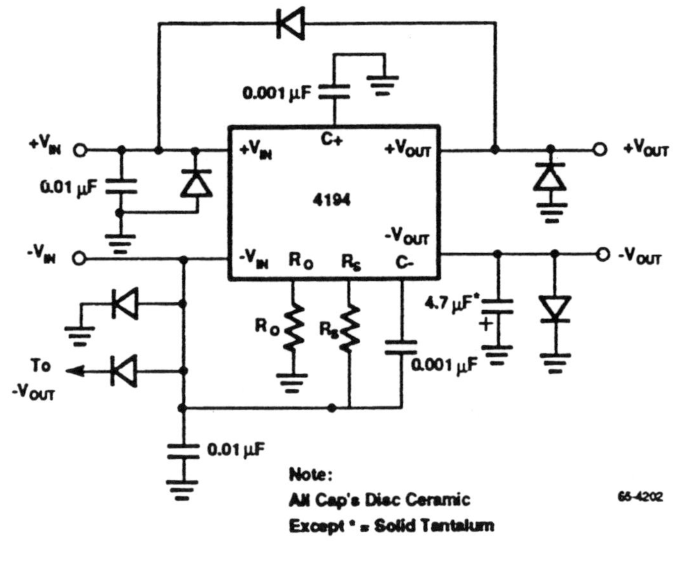

Figure 7–56. Diode protection points for dual-tracking regulator (*Raytheon Linear Integrated Circuits,* 1989, p. 9–72)

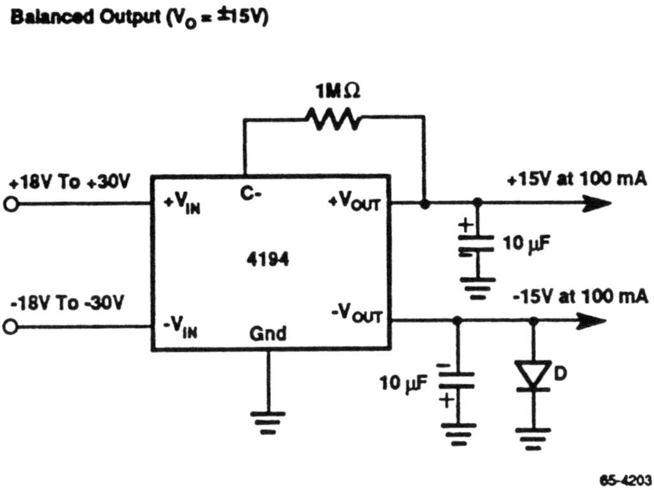

Figure 7–57. Brownout-protection circuit for dual-tracking regulator (*Raytheon Linear Integrated Circuits,* 1989, p. 9–72)

dropout voltage. When this happens, the negative output voltage can go positive. The maximum amount of current available is about 5 mA. In general, this is not enough current to damage most ICs that the regulator might be supplying, but it is a potentially destructive condition.

As shown in Fig. 7–57, a diode D can be connected to the negative output. If a small-signal silicon diode is used, the negative output voltage is clamped to about +0.55 V. A Schottky barrier or germanium diode clamps the voltage to about +0.3 V. The 1-M resistor connected between $+V_{OUT}$ and the C– terminal supplies drive to the negative output transistor, causing the transistor to saturate to $-V_{IN}$ during the brownout. Both the diode and resistor provide a simple way to keep the negative output negative (or at least at some low positive value) during any brownout condition.

7.12.5 Heat-Sinking

Although these regulators contain circuits that shut down the device should the temperature reach 175°C (as do most present-day linear-regulator ICs), such emergency shutdown should be avoided. The regulator will probably survive the shutdown, but not necessarily the circuits being supplied by the regulator. Some form of heat-sinking is the most practical means of keeping the junction temperature at a safe level; heat-sinking should always be considered when operating any linear regulator near the extremes of load current, ambient temperature, and input-output differential.

The following are examples of heat-sinking for the 4194 regulator. Compare these design examples to the basic heat-sink information in Chapter 2. Notice that the internal thermal-shutdown circuits start affecting load regulation at about 2°C below 175°C. However, the manufacturer recommends that design be based on a maximum junction temperature of 150°C to allow for process deviations.

In these examples, it is assumed that the basic equation for junction temperature is

$$T_J = T_A + P_D\,\theta_{JA},$$

where

T_J = junction temperature (°C)

T_A = ambient air temperature (°C)

P_D = power dissipated by device (W or mW)

θ_{JA} = thermal resistance from junction to ambient air (°C/W)

The basic equation can be rearranged to solve for P_D, or $P_D = (T_J - T_A)/\theta_{JA}$, and to solve for θ_{JA}, or $\theta_{JA} = (T_J - T_A)/P_D$.

The actual power dissipated by the regulator can be detailed as follows:

$$P_D = (V_{IN} - V_{OUT}) \times I_O + (V_{IN} \times I_Q),$$

where

V_{IN} = input voltage

V_{OUT} = regulated output voltage

I_O = output or load current

I_Q = quiescent current drain

This equation can be rearranged to find the actual output current, or $I_O = P_D - (V_{IN} \times I_Q)/(V_{IN} - V_{OUT})$.

Now let us combine these equations to find the output-current capability of the regulator (without a heat sink) for a given set of conditions. The following design factors are assumed:

T_J at thermal shutdown = 150°C (manufacturer's recommendation)

T_A = 125°C

θ_{JA} = 41.6°C/W; K (TO-66) package (see Table 7–9)

V_{IN} = 40 V

V_{OUT} = 30 V

I_Q = 1 mA + (75 μA/V_{OUT} × 30 V) = 3.25 mA*

*As shown in note 1 of Table 7–9, the current drain increases by 50 μA/V_{OUT} for positive and 10 μA/V_{OUT} for negative. The 75 μA/V_{OUT} is an average.

First, find the dissipation capability of the regulator at this high ambient temperature: $P_D = (T_J - T_A)/\theta_{JA}$, or (150 – 125)/41.6°C/W = 0.6 W or 600 mW.

Notice that this value can also be found if the regulator thermal-derating characteristics are given as shown in Table 7–9. (Derating factors are not always given for all linear regulators—as you will soon find out!)

As shown in Table 7–9, the maximum P_D at a T_A of less than 50°C is 2381 mW. Since our regulator must operate at a T_A of 125°C, derate at 23.81 mW/°C for the 75°C difference, or 75 × 23.81 = 1786; 2381 – 1786 = 595 mW, which is close to the 600 mW found using the other equation.

Finally, find the actual output-current capability, assuming that P_D is 600 mW (without a heat sink), or I_O = 600 – (40 × 3.25)/(40 – 30) = 470/10 = 47 mA. As a safety factor, subtract 10% of this value to find a realistic output current of about 42 mA.

Now assume that the regulator must provide 200 mA (0.2 A) of output current under the same operating conditions. Find the heat sink θ_{SA} °C/W rating that will dissipate the heat safely.

Notice that θ_{JA} represents the total junction-to-ambient thermal resistance. This is composed of the junction-to-case resistance, or θ_{JC} (found in Table 7–9 for our regulator), the case-to-heat-sink resistance, or θ_{CS} (heat sink compound, washer if any, etc.), and the heat-sink-to-ambient resistance or θ_{SA} (which is the heat-sink characteristic).

First find the P_D for our example, or P_D = (40 V – 30 V) × 0.2 A + (40 V × 0.00325 A) = 2.0 W + 0.13 W = 2.13 W. With a P_D of 2.13 W and a $T_J - T_A$ of 25°C, the total θ_{JA} is 11.75°C/W (25/2.13).

Table 7–9 shows a θ_{JC} of 7°C/W for our regulator in a TO–66 case, so there must be a maximum of 4.75°C/W between the case and ambient (11.75 – 7.0 = 4.75).

Allowing 0.5°C/W θ_{CS} for the heat-sink compound with metal-to-metal surfaces, a heat sink with 4°C/W θ_{SA} would be ideal, while 3 to 5°C/W θ_{SA} will probably do the job.

The heat-sink selection guide given in Chapter 2 shows that there are several readily available TO–66 heat sinks that meet the 3 to 5°C/W requirement, such as Wakefield 621, 623; Thermalloy 6606, 6129, 6141, 6303; IERC HP; and Staver V3-2-3.

7.12.6 Switchable Power Supply (On-Off)

Figure 7–58 shows how the outputs of our 4194 regulator can be switched simultaneously on or off under logic control. In the off state, the outputs are forced to a minimum voltage, or about ±20 mV, rather than becoming open circuit. The turn-on time, with the dual outputs programmed to ±12 V, is about 200 μs.

As shown in the internal schematic of Fig. 7–59, the switching circuit works by forcing the R_O pin to ground with an analog switch. A reference voltage that regulates with respect to $-V_S$ is generated at the R_S pin by zener Q_{12} and the buffer circuit of Q_{11} and Q_{13}. When the external 71.5-k R_S resistor is connected between the

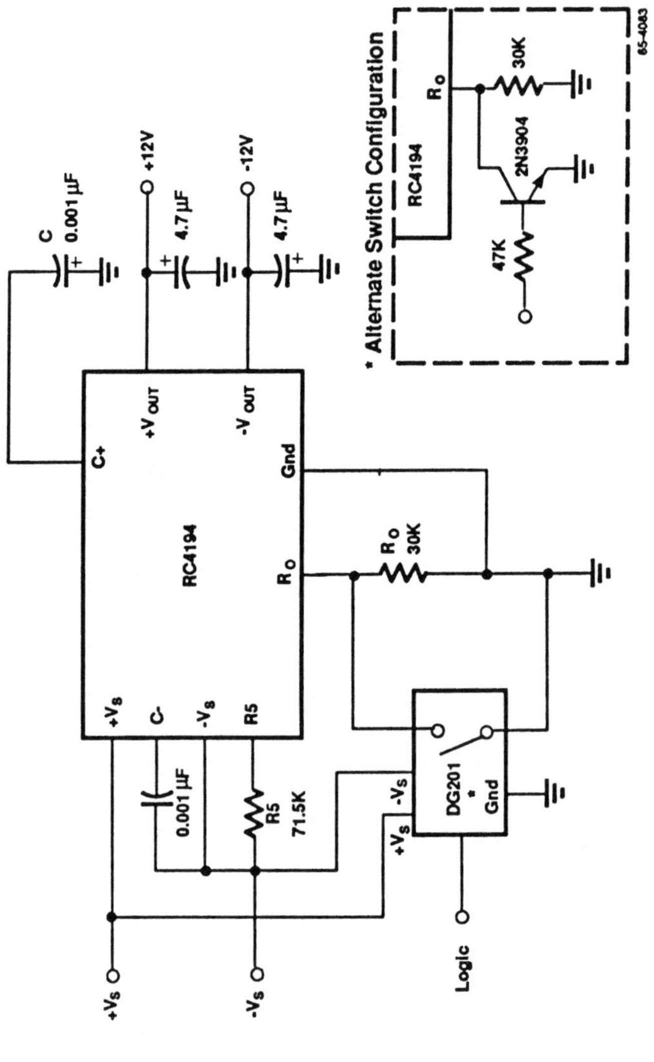

Figure 7–58. Switchable dual-tracking supply (*Raytheon Linear Integrated Circuits,* 1989, p. 9–70)

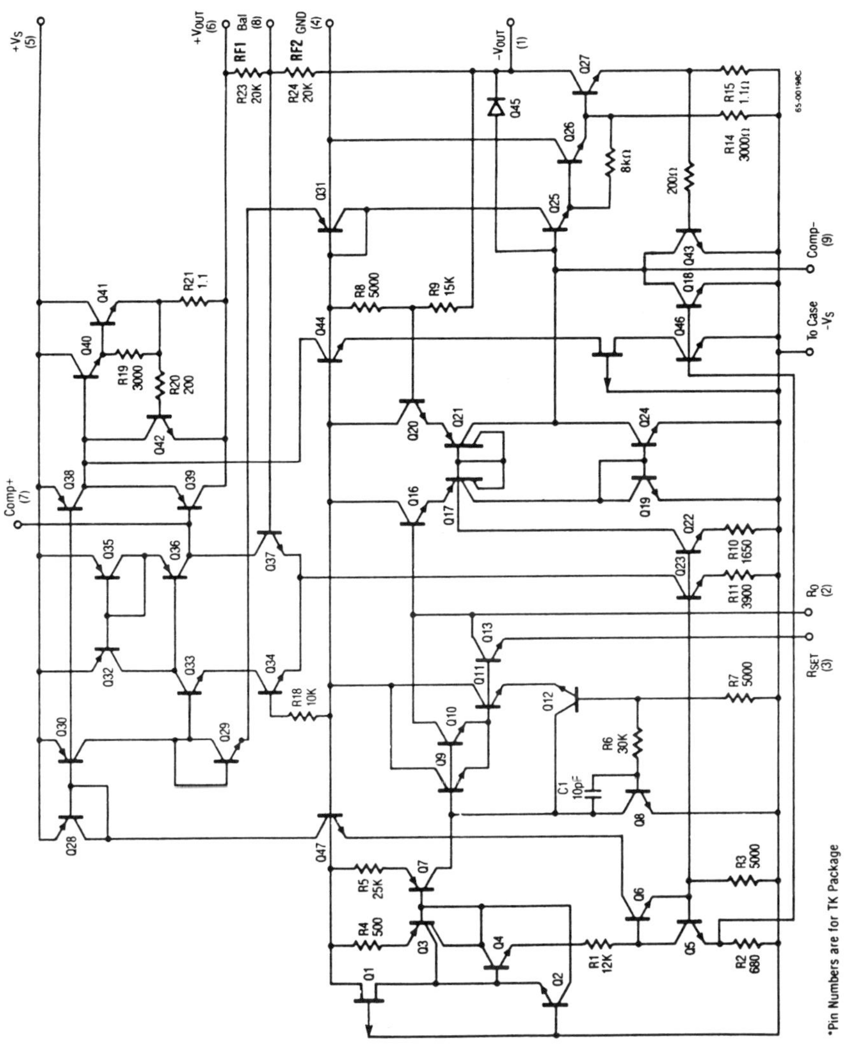

Figure 7–59. Internal schematic diagram of RC4194 dual-tracking regulator (*Raytheon Linear Integrated Circuits*, 1989, p. 9–76)

R_S pin and $-V_S$, a precision current of 100 μA current develops a ground-referenced voltage drop proportional to the value of R_O. The drop is then amplified by the internal error amplifier.

When the analog switch in Fig. 7–58 turns on, R_O is effectively shorted out. This causes 0 V to be applied at the error amplifier. In the off state, the output voltage is about ±20 mV. If a higher value (50 to 100 mV) is acceptable, then the DG201 FET switch shown can be replaced with a low-cost, small-signal transistor, as shown in the alternate configuration of Fig. 7–58.

7.12.7 Dual Supply with Higher Current Output

Figure 7–60 shows how the 4194 regulator can be connected to provide higher output currents. For the components shown, both the positive and negative outputs provide 2.5 A with a load regulation of 10 mV. Again, R_O sets the balanced output voltages. Notice that a value of 0.7 V is used to calculate the resistance of R_{SC}. Also, the filter capacitor at the positive output is optional, but not at the negative output. However, any capacitors used must be as close to the IC as possible, preferably at the IC pins.

7.12.8 Dual Supply for Multiple Op-Amps

Figure 7–61 shows how the RM4194 (military version) regulator can be connected to supply many op-amps. For the components shown, a single regulator can supply about 180 741 op-amps (or equivalent). The value of R_O sets the balanced out-

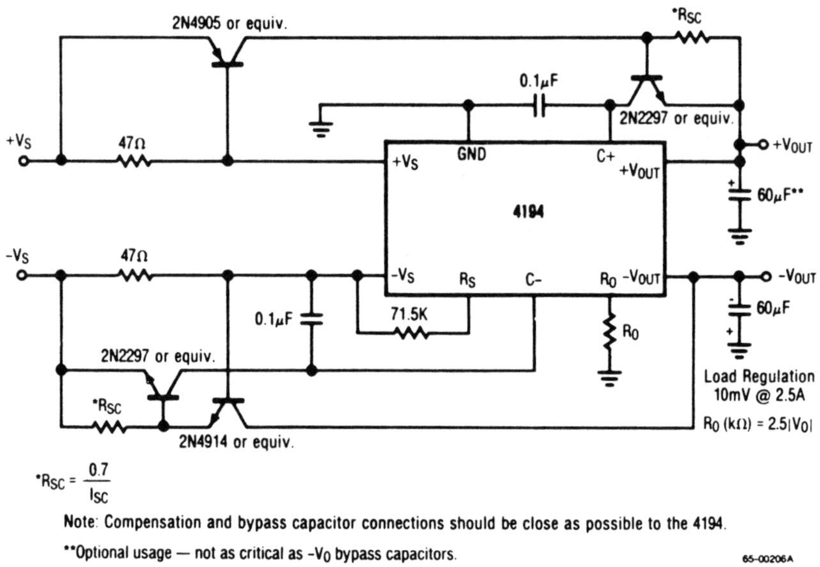

Figure 7–60. Dual supply with higher current output (*Raytheon Linear Integrated Circuits,* 1989, p. 9–68)

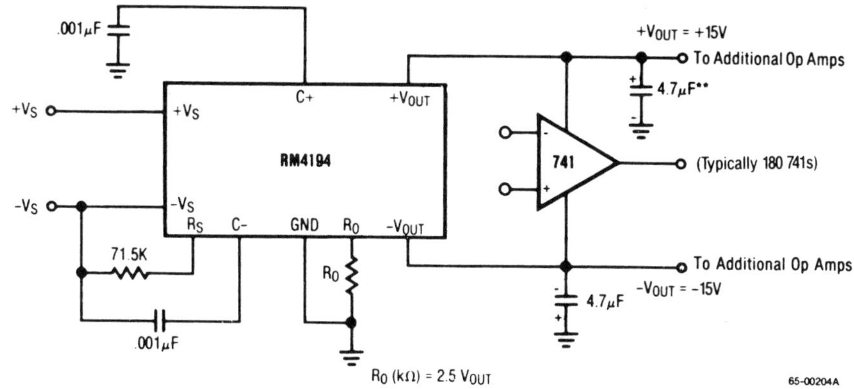

Figure 7–61. Dual supply for multiple op-amps (*Raytheon Linear Integrated Circuits,* 1989, p. 9–69)

put voltages. Again, the positive-output filter capacitor is optional (but not the negative output), and always keep the capacitors as close to the IC pins as is practical.

7.12.9 Digitally Controlled Dual 200-mA Voltage Regulator

Figure 7–62 shows how the RM4194TK regulator can be connected to provide digitally controlled dual output voltages. Notice that the scale-factor resistor R_S is not used, and there is no connection to the R_S pin. Also, the output-voltage resistor R_O is replaced by the digital-to-analog converter DAD–08 and resistor R_1. Follow the adjustment notes for proper output voltage. Notice that R_2 is adjusted for –19.92 V and $-V_{OUT}$ when the logic input to the DAC–08 is 11111111 (all of the binary inputs tied to +5 V). Then R_3 is adjusted for +19.92 V at $+V_{OUT}$, for best tracking.

7.12.10 Unbalanced Output Voltages

Figure 7–63 shows how the 4194 regulator can provide unbalanced dual outputs (to 38 RM710 comparators in this example). The scale-factor resistor R_S is still 71.5 k, and R_O is selected for –6 V at $-V_{OUT}$ (use 15 k for R_O, 2.5 k × 6 = 15 k). The relationship between $-V_{OUT}$ and $+V_{OUT}$ is set by R_A and R_B, as shown by the note. (The schematic referred to in the note is Fig. 7–59.) For a $+V_{OUT}$ of 12 V, when $-V_{OUT}$ is –6 V, omit R_A and use 20 k for R_B.

7.13 Battery-Charging Systems

Linear supplies are well suited for battery-charger applications. This section describes four such circuits using the classic National Semiconductor LM317 three-

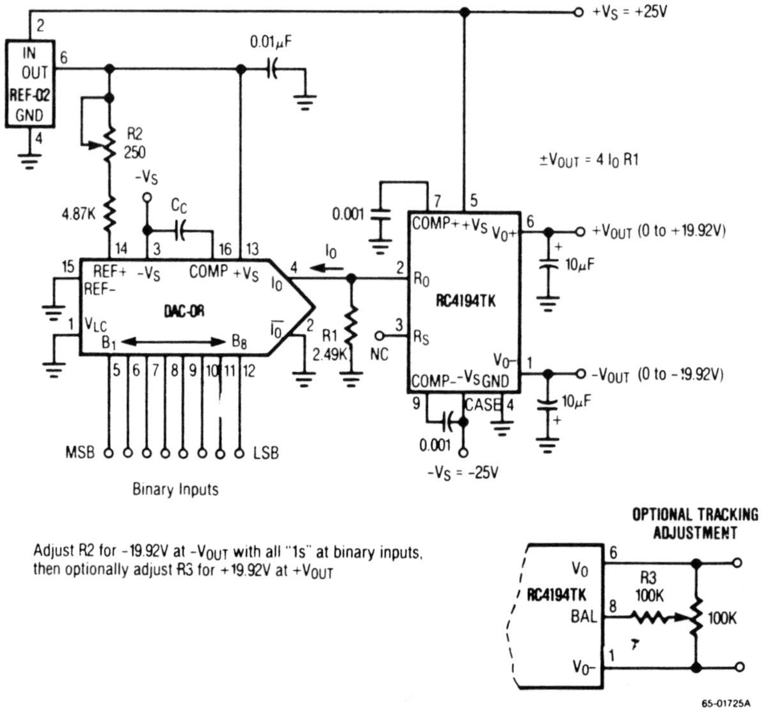

Figure 7–62. Digitally controlled dual 200-mA voltage regulator (*Raytheon Linear Integrated Circuits,* 1989, p. 9–69)

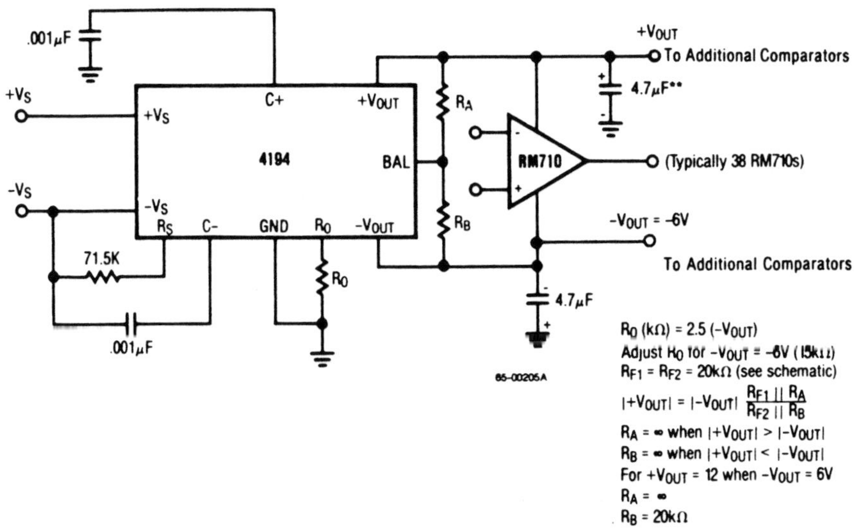

Figure 7–63. Dual supply with unbalanced outputs (*Raytheon Linear Integrated Circuits,* 1989, p. 9–68)

terminal regulator. The LM317 is especially attractive for constant-voltage applications because of the ability to accurately adjust the output voltage. Constant-voltage charging circuits provide the fastest way to charge batteries. However, close control of the charging voltage is necessary to prevent overcharging, especially with nickel-cadmium cells. The internal-protection circuits of the LM317 protect against accidental overload conditions commonly found in charging systems.

7.13.1 Basic Constant-Voltage Charger

Figure 7–64 shows the LM317 connected in the basic constant-voltage charger circuit. The output voltage is set with resistors R_2 and R_3 and given by $V_{OUT} = 1.25(1 + R_3/R_2)$. The peak charging current or output current is controlled by the internal current limit of the LM317. This current limit works even if a battery is connected backwards to the charger output. The thermal-limiting circuits decrease the output current, protecting the regulator as well as the transformer, should a fault condition exist for an extended period of time.

The peak output voltage from the transformer must be at least 50% greater than the battery voltage, plus 3 V. This is because there are no filter capacitors at the rectifier output, and the battery is charged only on the sine-wave peaks. However, the average current from the transformer is lower than with capacitive-input filter circuits, so there should be no significant increase in cost.

Resistors R_1 and R_4 provide additional control of the charging characteristics. R_1 sets the output impedance of the circuit, allowing a taper-charge characteristic to be generated. The LM317 can also be used to limit the peak-charging current (to a partially charged battery) at a value other than the LM317. With R_1 in the circuit, the output impedance is $Z_{OUT} = R_1\ (1 + R_3/R_2)$. With R_1 in the circuit, the value of

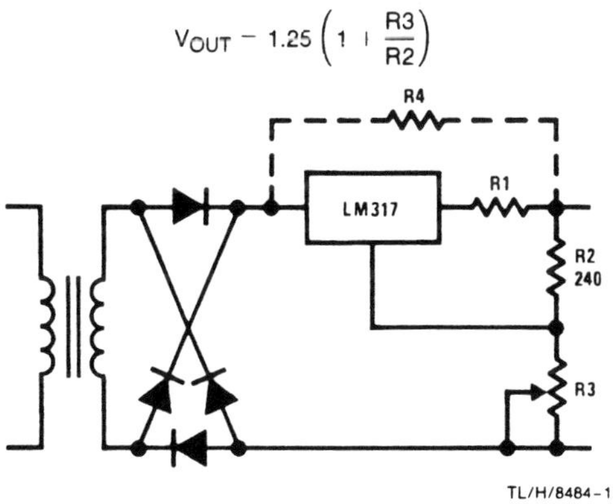

Figure 7–64. Constant-voltage battery charger (National Semiconductor, *Linear Applications Handbook,* 1991, p. 1239)

the resistor needed for a particular output impedance is decreased, reducing cost and power dissipation.

For example, with a 6-V gelled-electrolyte battery, the LM317 can be set to give a 6.9-V output by making R_1 0.4 ohm. This limits the charging current to 0.5 A at the start of charging (rather than the internal current limit of the LM317). With the 0.4-ohm value, the peak output current is about 2 A, with R_1 dissipating 1.6 A (in contrast to a dissipation of about 8 W if R_1 is 2 ohms).

With resistor R_4 in the circuit, there is no damage if the input power is removed, and a fully charged battery is connected to the charger output. Under these conditions, about 5 mA of current is drawn by the divider (R_2, R_3). Because there is no ground connection to the LM317, very little current flows through the LM317. Notice that the LM317 differs from some other three-terminal regulators in that there is no damage if power is applied to the output terminal with the input open-circuited. If the battery is connected backwards, the LM317 will current-limit and thermal-limit normally, protecting the charger.

7.13.2 Constant-Voltage Charger with Peak-Current Limiting

Figure 7–65 shows an improvement on the basic circuit to provide peak-current limiting. In this circuit, R_1 senses the output current and turns on Q_1 when $I_{OUT} \times R_1$ equals about 0.6 V. Transistor Q_1 pulls the adjustment terminal of the LM317 negative, decreasing the output voltage and controlling the output current. One problem with this circuit is that it does not work for direct short circuits. The output voltage must be above about 0.6 V for the external current limiting to be active.

Of course, the internal current limit of the LM317 is still operative. And because most batteries charge to above 0.6 V very quickly, the problem should not be

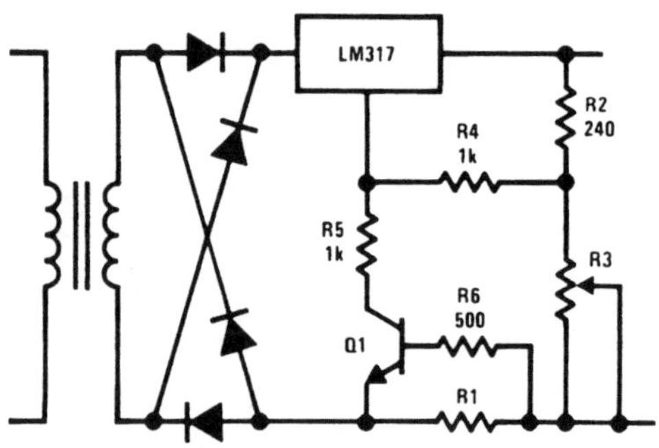

Figure 7–65. Constant-voltage battery charger with peak-current limiting (National Semiconductor, *Linear Applications Handbook,* 1991, p. 1239)

of real concern. Resistors R_4, R_5, and R_6 protect transistor Q_1 for both direct short circuits and reverse-battery connection.

7.13.3 No Battery Loading with Power Off

Figure 7–66 shows a circuit where all loading is removed from the battery when input power is off. Such a circuit is best suited for standby applications. When power is on, Q_1 is saturated, grounding the voltage-setting divider R_2, R_3. Under these conditions, the circuit is similar to that of Fig. 7–64.

When power is off, Q_1 is open, eliminating any loading on the battery. A separate pair of low-current diodes D_1 and D_2 are necessary to bias Q_1, rather than the power bridge rectifiers. If R_1 was tied to the output of the bridge, reverse current flow through the LM317 would keep Q_1 on, and loading the battery.

7.13.4 Constant-Current Charger

Figure 7–67 shows the LM317 connected as a constant-current charger. Resistor R_1 between the adjustment terminal and the output sets the output current at $I_{OUT} = 1.25/R_1$. Current can be set at any point between about 10 mA and 1.5 A with R_1. Current regulation is very tight at any current level because only 50 μA flows out of the adjustment terminal. The circuit is also immune to damage from shorts or

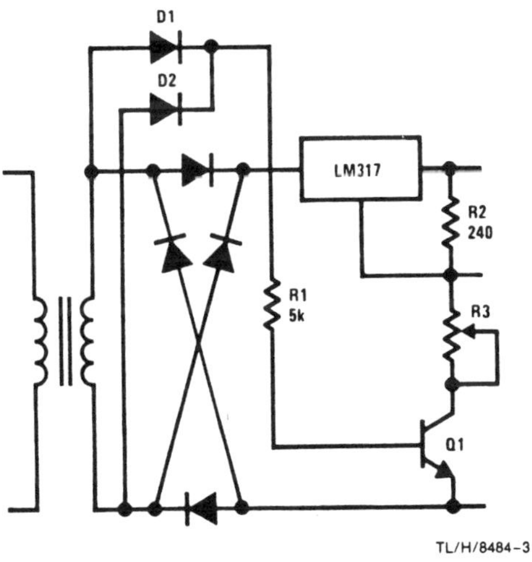

Figure 7–66. Battery charger with loading removed when power is off (National Semiconductor, *Linear Applications Handbook,* 1991, p. 1240)

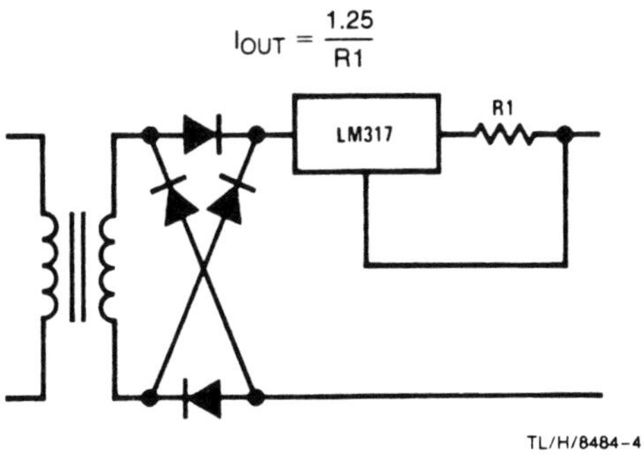

Figure 7–67. Constant-current battery charger (National Semiconductor, *Linear Applications Handbook,* 1991, p. 1240)

reverse-battery connections. The input voltage should be about 1.5 times the battery voltage, plus 3 V.

7.14 Three-Terminal Regulators with Kelvin Sensing and Parallel Capability

This section describes two circuits to add remote sensing to a three-terminal regulator. One of the two circuits also adds parallel capability. Generally, parallel operation of three-terminal regulators is not recommended. This is because the regulators do not share current equally. For example, if you try to make a 3-A regulator using three 1-A regulators, the device with the highest output could be carrying 2.5 A in a current-limit mode. The regulator with the second-highest output carries only 0.5 A, while the third regulator might be completely off.

It is obvious that such a system is highly unreliable. A simple way to improve current sharing is to insert a low-value resistor in series with the output of each regulator. Unfortunately, if the resistors are made large enough to provide good sharing, load regulation is impaired. The circuits in this section use remote sensing (see Section 6.10) to overcome the problems of a high resistance value in the output.

7.14.1 Kelvin or Remote Sensing

Figure 7–68 shows a remote-sensing scheme using a National Semiconductor LM338 5-A adjustable regulator. A_1 forces a voltage drop across R_3 equal to the voltage across the parasitic resistance r_s. The current through R_3 flows into the output of A_1 and out the negative-supply terminal (pin 4). This creates a voltage drop across R_4 just equal to the voltage across r_s, canceling the effect of r_s on load regulation. There is an error in V_{OUT} created by the quiescent current of R_1. However, for a 5-V output, the error is only about 0.7%.

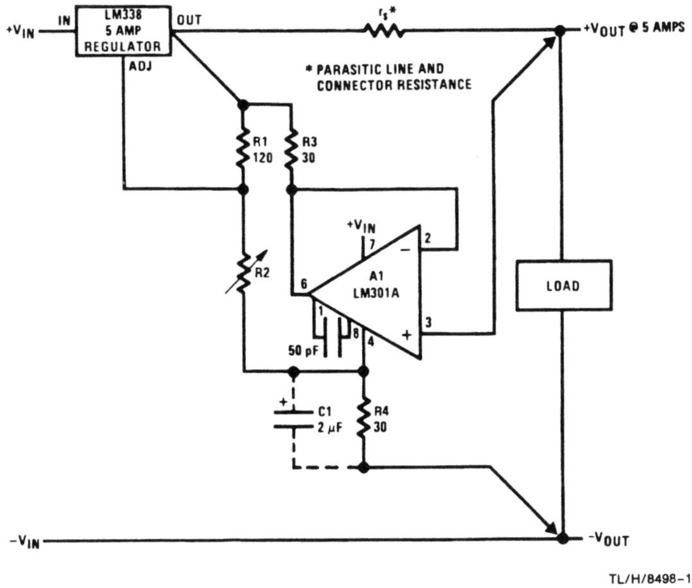

Figure 7–68. Kelvin or remote sensing (National Semiconductor, *Linear Applications Handbook,* 1991, p. 1263)

The voltage loss across r_s must be limited to about 300 mV to avoid current limiting in A_1. If a larger drop must be accommodated (say, when the regulator must be on a different card from the load), both R_3 and R_4 must be increased. Capacitor C_1 is necessary only if intermediate values of capacitance (2 µF – 20 µF) are put directly across the load (for filtering or bypass). Noticc that other positive regulators may be used in place of the LM338.

7.14.2 Parallel Operation with Remote Sensing

Figure 7–69 combines the sensing with parallel operation, where the voltage across the current-sharing resistors is corrected by the sensing connection. r_{s1} through r_{s3} are equal lengths of No. 22 lead wire, which act as ballasting resistors. These resistors can be kept small because the LM338 adjustment pins are paralleled, forcing the outputs to track to within about 60 mV.

Resistance r_{s4} consists of the parasitic resistance of any additional output lead plus connector loss. The total loss for r_{s4} may be up to 0.25 V without loss of proper sensing.

Notice that if U_1 has the lowest reference voltage of the three regulators, full sensing might not become effective until the output current has increased above a

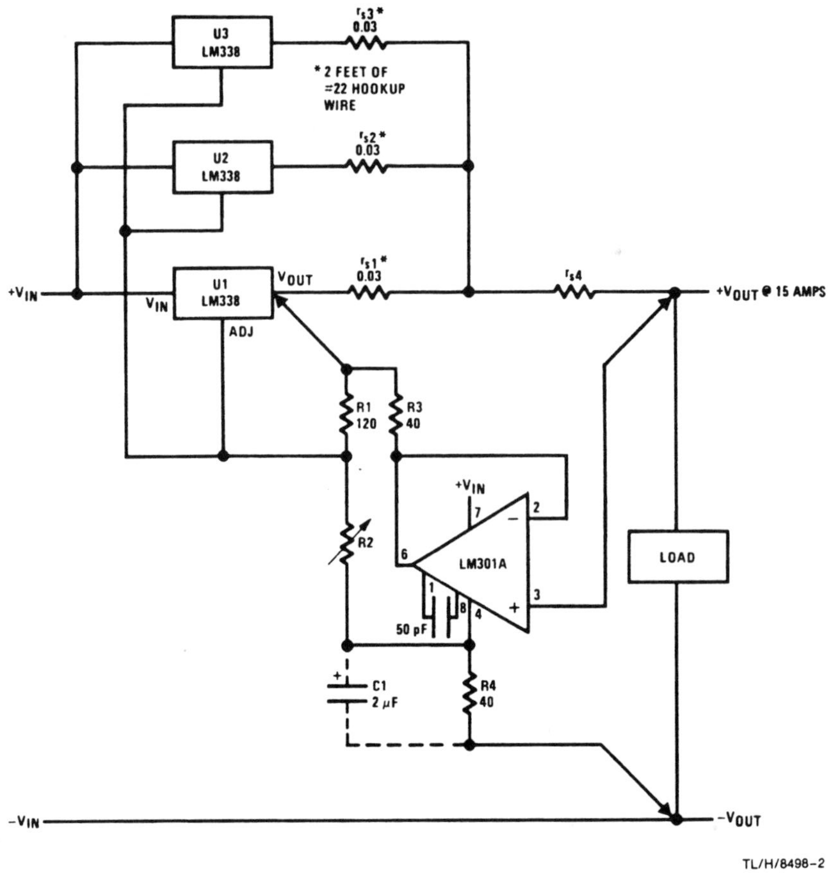

Figure 7–69. Parallel operation with remote sensing (National Semiconductor, *Linear Applications Handbook,* 1991, p. 1264)

threshold value of several amperes. If this is undesirable, the adjustment pin of U_1 may be connected to a 5-ohm tap on R_1. This increases the effective reference voltage by about 50 mV. The current load for U_1 would be 1.5 A higher, however.

7.15 Low-Drift Voltage References

This section describes some low-drift voltage-reference circuits using four different approaches. These circuits were originally developed by National Semiconductor as low-drift references for analog/digital converters where precision voltages are required. The first circuit uses a low-cost zener but precision low-drift resistors to get the desired drift. The second circuit uses standard 1% resistors but a precision temperature-stabilized zener. The third circuit is for low-voltage outputs (where the output must be less than the zener voltage). The fourth circuit provides ultra-low-drift reference voltages.

7.15.1 Voltage Reference with Low-Cost Zener

Figure 7–70 shows an op-amp/zener circuit (see Chapter 5), where an LM308 op-amp is used to increase the zener output to 10 V, while adding a worst-case drift of 4 ppm/°C to the 10 ppm/°C of the zener. Resistors R_3 and R_4 do better than 10 ppm, bringing the total error to 18 ppm. Potentiometer R_5 and resistor R_2 are added to eliminate the initial zener tolerance. The loading on R_5 by R_2 is small, and there is no tracking requirement between R_5 and R_2. However, it is necessary for R_2 to track R_3 and R_4 within 50 ppm to maintain the 20-ppm/°C 10-V output.

7.15.2 Voltage Reference with Standard 1% Resistors

Figure 7–71 shows a temperature-stabilized zener (National Semiconductor LM399) and an op-amp used to produce a total drift (exclusive of resistors) at 3 ppm/°C. With this circuit, the resistor-tracking requirement is relaxed to about 50 ppm, allowing standard 1% resistors to be used.

The circuit can be modified easily (for applications requiring 3 to 5 ppm/°C overall drift) by tightening the tracking of the resistors. The Kelvin sensing for both output and ground provides additional precision. For even lower drifts, substitute a

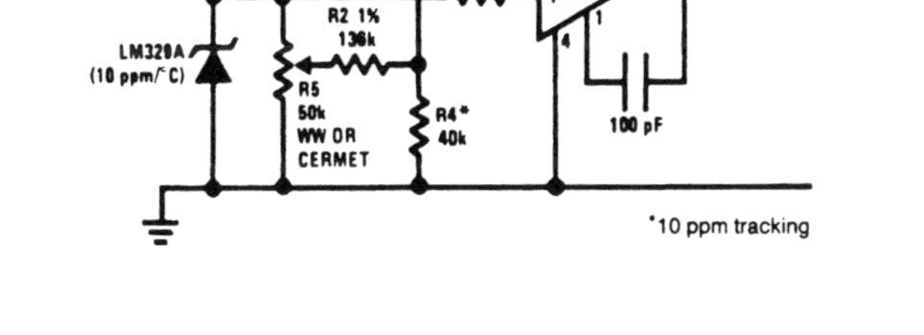

TL/H/5615-1

Figure 7–70. Low-drift voltage reference with low-cost zeners (National Semiconductor, *Linear Applications Handbook,* 1991, p. 458)

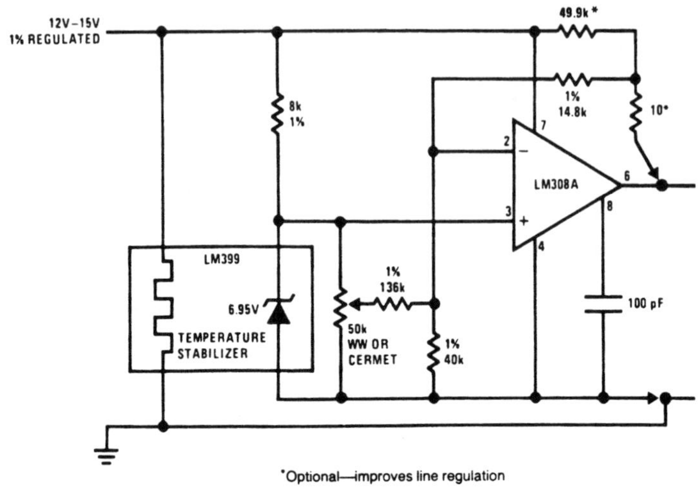

Figure 7–71. Low-drift voltage reference with 1% resistors (National Semiconductor, *Linear Applications Handbook,* 1991, p. 459)

1-μV/°C op-amp, 1-ppm tracking resistors, and an LM199A zener. Such a circuit can provide overall drifts of 1 ppm/°C.

7.15.3 Low-Voltage Reference

Figure 7–72 shows how an op-amp and zener can be combined to produce a reference where the output voltage is less than the zener voltage (a 5-V reference

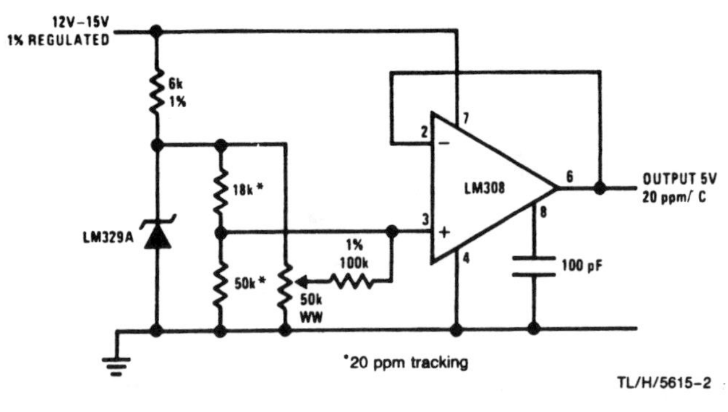

Figure 7–72. Low-drift voltage reference (National Semiconductor, *Linear Applications Handbook,* 1991, p. 459)

Table 7–10. Drift error contribution from reference components (National Semiconductor, *Linear Applications Handbook,* 1991, p. 458)

DEVICE	ERROR	10V OUTPUT DRIFT
Zener	**Zener Drift**	
LM199A	0.5 ppm/°C	0.5 ppm/°C
LM199, LM399A	1 ppm/°C	1 ppm/°C
LM399	2 ppm/°C	2 ppm/°C
1N829, LM3999	5 ppm/°C	5 ppm/°C
LM129, 1N823A, 1N827A, LM329A	10–50 ppm/°C	10–50 ppm/°C
LM329, 1N821, 1N825	20–100 ppm/°C	20–100 ppm°C
Op Amp	**Offset Voltage Drift**	
LM725, LH0044, LM121	1 μV/°C	0.15 ppm/°C
LM108A, LM208A, LM308A	5 μV/°C	0.7 ppm/°C
LM741, LM101A	15 μV/°C	2 ppm/°C
LM741C, LM301A, LM308	30 μV/°C	4 ppm/°C
Resistors	**Resistance Ratio Drift**	
1% (RN55D)	50–100 ppm	20–40 ppm/°C
0.1% (Wirewound)	5–10	2–4 ppm
Tracking 1 ppm Film or Wirewound	—	0.4 ppm/°C

with a 15-V input). In this case, zener drift contributes proportionally to the output drift, while op-amp offset drift adds at a greater rate. Table 7–10 shows the drift-error contribution of the three principal components (zener, op-amp, resistors) in a zener/op-amp voltage-reference or regulator circuit.

7.15.4 Ultra-Low-Drift Voltage Reference (with TC Trimming)

Figure 7–73 shows a circuit for temperature-coefficient (TC) trimming. Such trimming is necessary to get the lowest possible drift. With discrete zeners, the operating current can sometimes be trimmed to change the reference TC. However, the TC is not always linear or predictable. With IC zeners, TC is independent of operating current, so trimming must be done elsewhere in the circuit. The lowest-drift components should be used because trimming can remove only a linear component of drift. High-TC devices can have a highly nonlinear drift, making trimming difficult.

In the circuit of Fig. 7–73, an LM199A reference with 0.5 ppm/°C is used with a 121/108 op-amp combination. Resistors should be 1-ppm tracking to give overall untrimmed drifts of about 0.9 ppm. The 121/108 is a low-drift amplifier combination where drift is predictably proportional to offset voltage. An offset can be set for the 121/108 combination to cancel the measured drift as follows.

Disconnect the zener and ground the op-amp input. Null the op-amp offset to zero. Reconnect the zener and adjust the output to precisely 10 V. Make a temperature run and note the drift. Typically the op-amp will drift 3.6 μV/°C for every 1 mV of offset. So, for every 5-μV/°C drift at the output, the offset of the op-amp is adjusted 1 mV (1.4 mV measured at the output) in the opposite direction. Then readjust the output to 10 V and check the drift.

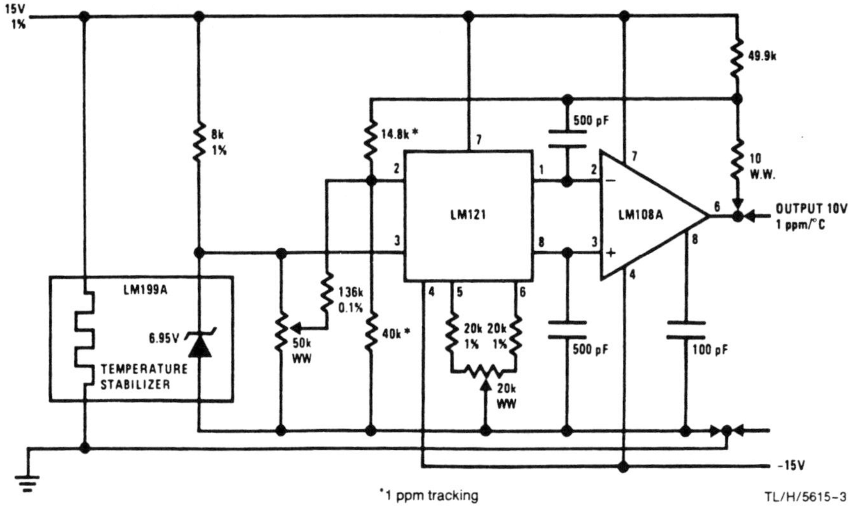

Figure 7–73. Ultra-low-drift voltage reference (National Semiconductor, *Linear Applications Handbook,* 1991, p. 460)

Notice that oven testing can sometimes cause thermal gradients across circuits, giving 50 to 100 μV of error. However, with careful layout and trimming, overall reference drifts of 0.1 to 0.2 ppm/°C can be achieved.

There are two other possible problem areas to be considered before final layout. Good single-point grounding is important. Traces on a PC board can easily have 0.1 ohm, and only 10 mA will cause a 1-mV shift. Also, because these circuits are designed for use with high-speed digital equipment, shielding may be necessary to prevent pickup at the op-amp inputs. Transient response to pickup, or rapid load changes, can sometimes be improved by a large capacitor (1 to 10 μF) directly on the op-amp output. Of course, this depends on op-amp stability.

7.16 Additional Three-Terminal-Regulator Circuits

This section describes a number of circuits based on the use of a three-terminal regulator (the classic National Semiconductor LM340). These circuits are complete, including input/output characteristics and component values, so the circuits can be used immediately as is. The discussions in the following paragraphs are provided to help readers adapt the circuits to their specific needs.

7.16.1 High-Current Regulator with Short-Circuit Current Limit

Figure 7–74 shows the LM340 connected to provide high current output. This 15-V regulator includes an external series-pass transistor Q_1 to increase the output

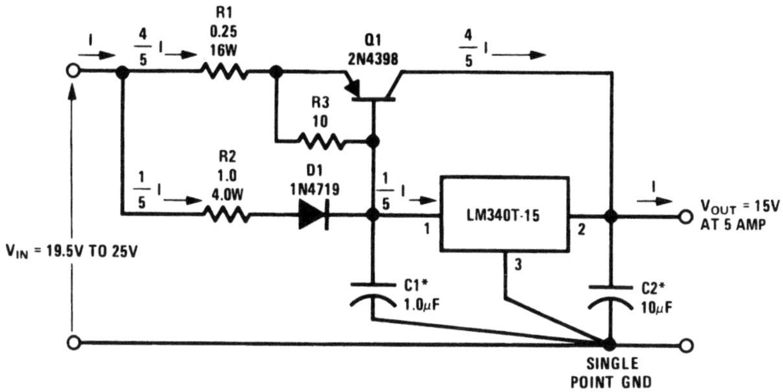

*Solid tantalum

Note 1: Current sharing between the LM340 and Q1 allows the extension of short circuit current limit, safe operating area protection, and (assuming Q1's heat sink has four or more times the capacity of the LM340 head sink) thermal shutdown protection.

Note 2: $I_{SHORT\ CIRCUIT}$ is approximately 5.5 amp.

Note 3: $I_{OUT\ MAX}$ at V_{OUT} = 15V is approximately 9.5 amp. TL/H/7413–6

Figure 7–74. High-current regulator with short-circuit current limit (National Semiconductor, *Linear Applications Handbook,* 1991, p. 345)

current capability to 5 A. Unlike most current-boosting methods, this circuit maintains the LM340's ability to provide short-circuit current limiting and thermal shutdown without additional active components. The extension of these safety features to the series-pass transistor Q_1 is based on a current-sharing scheme using R_1, R_2, and D_1.

Assuming that the base-to-emitter voltage of Q_1 and the voltage drop across D_1 are equal, the voltage drops across R_1 and R_2 are equal. The currents through R_1 and R_2 will then be inversely proportional to their resistances. For the example shown in Fig. 7–74, resistor R_1 will have four times the current flow of R_2. For reasonable values of Q_1 beta, the current through R_1 is approximately equal to the collector current of Q_1. The current through R_2 is equal to the current flowing through the LM340. Therefore, under short-circuit or overload conditions, the protection circuits of the LM340 will limit its own output current and, because of the R_1/R_2 current-sharing scheme, the output current of Q_1 as well.

Thermal-overload protection also extends to Q_1 when the Q_1 heat sink has four or more times the capacity of the LM340 heat sink. This follows from the fact that both devices have approximately the same input/output voltage and share the load current in the ratio of four to one (as shown by the current arrows in Fig. 7–74).

The circuit normally operates at output currents up to 5 A. This means that up to 1 A of current flows through the LM340 and up to 4 A flows through Q_1. For short-term overload conditions, the curve of Fig. 7–75 shows the maximum instantaneous output current versus temperature for the circuit of Fig. 7–74. This curve re-

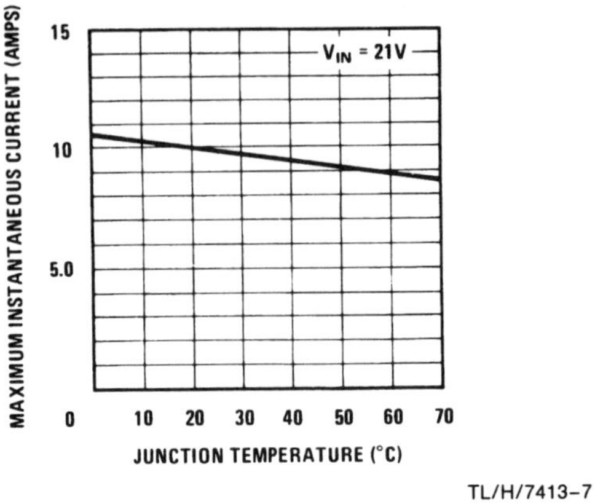

Figure 7–75. Maximum instantaneous current versus junction temperature for high-current regulator (National Semiconductor, *Linear Applications Handbook,* 1991, p. 345)

flects the approximate 2-A current limit of the LM340, causing an 8-A current limit in the series-pass transistor Q_1, or 10 A total.

Under continuous short-circuit conditions, the LM340 will heat up and limit to a steady total-state short-circuit current of 4 to 6 A, as shown in Fig. 7–76. This curve was taken using a Wakefield 680–75 heat sink (approximately 7.5°C/W) at a 25°C ambient temperature.

For optimum current sharing (over temperature) between the LM340 and Q_1, the diode D_1 should be physically located close to Q_1 on the heat sink. This should be done in such a manner that D_1 and Q_1 are kept at the same temperature. The

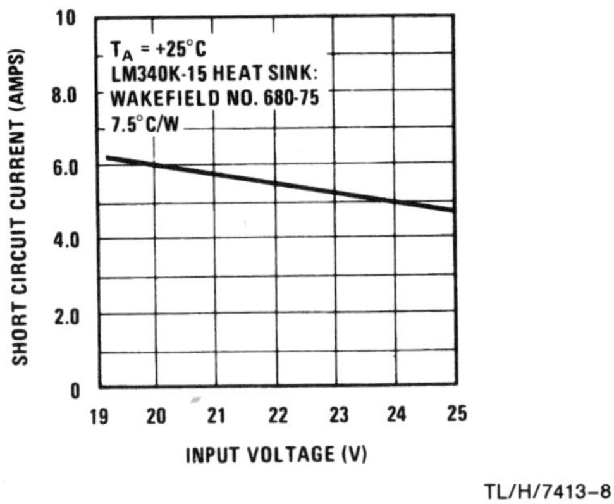

Figure 7–76. Continuous short-circuit current versus input voltage for high-current regulator (National Semiconductor, *Linear Applications Handbook,* 1991, p. 345)

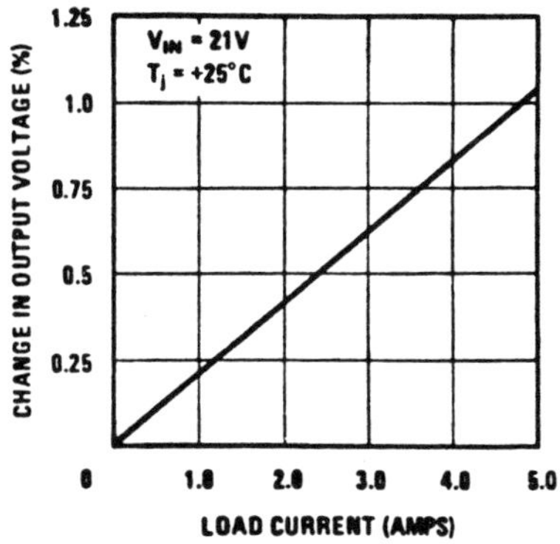

Figure 7–77. Load-regulation versus load-current for high-current regulator (National Semiconductor, *Linear Applications Handbook,* 1991, p. 346)

LM340 must be electrically isolated from the heat sink because the case (pin 3) is at ground potential and the Q_1 case (collector) is at the output potential of the circuit.

Capacitors C_1 and C_2 are required to prevent oscillation and improve the output impedance, respectively. Resistor R_3 provides a path to unload excessive base charge from the base of Q_1 when the regulator goes suddenly from full load to no load.

The single-point ground system (see Section 6.10) allows the sense pins (2 and 3) of the LM340 to monitor the voltage directly at the load rather than at some point along a (possibly) resistive ground-return line carrying up to 5 A of load current.

Figure 7–77 shows the typical variation of load regulation versus load current for the circuit. The insertion of the series-pass transistor Q_1 increases the input/output differential voltage from 2 V to approximately 4.5 V. For an output current less than 5 A, the R_1/R_2 ratio can be set lower than 4:1. Therefore, a less expensive PNP transistor may be used.

7.16.2 TTL Voltage Regulator

Figure 7–78 shows the LM340 connected to provide high current output for TTL circuits. This 5-V regulator uses a relatively inexpensive NPN series-pass transistor Q_1 with a lower power PNP transistor Q_1 to replace the single, higher cost, higher power PNP used in the circuit of Fig. 7–74.

The circuit of Fig. 7–78 provides a 5-V output at up to 5 A of load current with a typical load regulation of 1.8% from no load to full load. During test, the peak instantaneous output current observed was 10.4 A at a 25°C junction temperature (pulsed load with a 1 ms on and 200 ms off period) and 8.4 A for a continuous short circuit. The typical line regulation is 0.02% of input voltage change ($I_{OUT} = 0$).

Diode D_3 (a National Semiconductor NSL5027 LED) can be added as an overload indicator. This is shown in dotted lines on Fig. 7–78. With the overload-

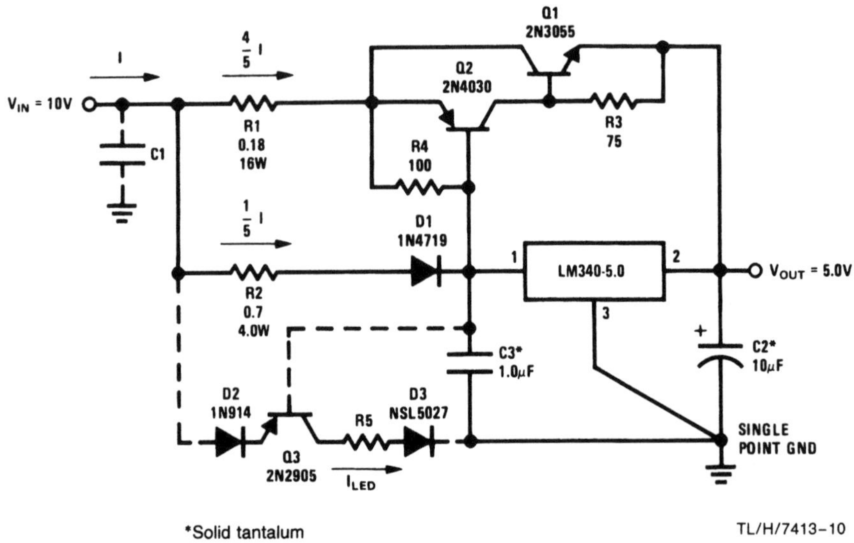

Figure 7–78. TTL voltage regulator (National Semiconductor, *Linear Applications Handbook,* 1991, p. 346)

indicator configuration, R_2 is not only a current-sharing resistor but also an overload sensor. Resistor R_5 determines the current through the LED. Diode D_2 is added to match the drop across D_1.

Once the load current exceeds 5 A (1 A through the LM340, assuming perfect current sharing and $V_{D_1} = V_{D_2}$), Q_3 turns on, as does the overload LED D_3.

7.16.3 Adjustable-Output Regulator for Intermediate Voltages

Figure 7–79 shows the LM340 connected to provide a nonstandard output voltage while maintaining the built-in limiting features. The circuit provides a 10-V output using an LM304K-5.0 by raising the reference (pin 3) of the regulator by 5 V.

The 5-V pedestal results from the sum of regulator quiescent current I_Q and a current equal to V_{REG}/R_1, flowing through potentiometer R_2 to ground. R_2 is made adjustable to compensate for differences in I_Q and V_{REG} output. The circuit is practical because the change in I_Q (caused by line-voltage and load-current changes) is quite small.

Line regulation for the circuit is the sum of the LM340 line regulation, the effect of current through R_2, and the effects of the difference in I_Q (ΔI_Q) in response to input-voltage changes. The change in output voltage is

$$V_{OUT} = (Lr/V)\Delta V_{IN} + \frac{(Lr/V)\Delta V_{IN} R_2}{R_1} + (QCC/V)\Delta V_{IN} R_2,$$

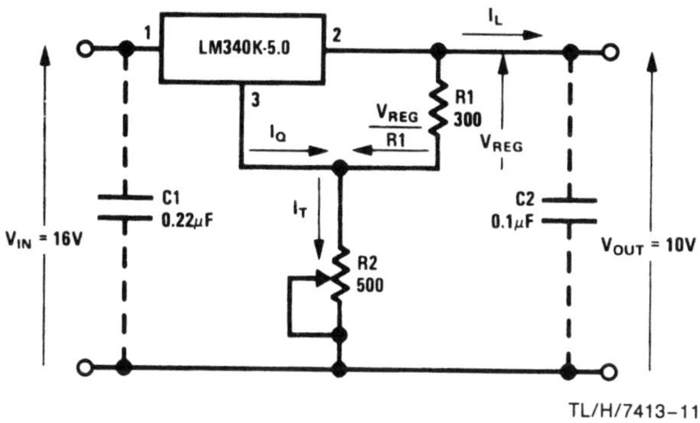

Figure 7–79. Adjustable-output regulator for intermediate voltages (National Semiconductor, *Linear Applications Handbook,* 1991, p. 346)

where

Lr/V = line regulation per volt; the change in the LM340 output voltage per volt of input/output voltage change at a given I_{OUT}

QCC/V = Quiescent current change per volt of input/output (pin 1 to pin 2) voltage change of the LM340.

This gives a total line regulation of

$$\frac{\Delta V_{OUT}}{\Delta V_{IN}} = (\mathrm{Lr/V})(1 + R_2/R_1) + (\mathrm{QCC/V})\, R_2$$

The LM340-5.0 data sheet lists $\Delta V_{OUT} < 50$ mV and Δ50 mV and $\Delta I_Q < 1$ mA for $\Delta V_{IN} = 18$ V at $I_{OUT} = 500$ mA. This is

$$\mathrm{Lr/V} = (50\ \mathrm{V}/18\ \mathrm{V}) \approx 3\ \mathrm{mV/V}$$

$$\mathrm{QCC/V} = (1\ \mathrm{mA}/18\ \mathrm{V}) = 55\ \mu\mathrm{A/V}$$

Using these values to find worst-case line regulation (with an I_{OUT} of 500 mA and R_2 at 310 ohms) is

$$\frac{\Delta V_{OUT}}{1\ \mathrm{V}} = 3\ \mathrm{mV/V}\ (1 + 310/300) + (55\ \mu\mathrm{A/V})\ 310$$

$$\frac{\Delta V_{OUT}}{1\ \mathrm{V}} = 6\ \mathrm{mV/V} + 17\ \mathrm{mV/V} = 23\ \mathrm{mV/V}$$

This represents a worst-case line regulation of 0.23%/V.

The load regulation is the sum of the LM340 voltage regulation, the effect of current through R_2, and the effect of ΔI_Q in response to changes in load current. Using the following definitions and the notation shown in Fig. 7–79, ΔV_{OUT} is

Z_{OUT} = regulator output impedance; the change in output voltage per amp of load-current change

Z_{340} = the LM340 output impedance

QCC/A = Quiescent current change per amp of load current change.

$$V_{OUT} = (Z_{340})\Delta I_L + (Z_{340})/R_1\ \Delta I_L\ R_2 + (QCC/A)\Delta I_L\ R_2$$

and the total output impedance is:

$$Z_{OUT} + \frac{\Delta V_{OUT}}{\Delta I_L} = Z_{340}(1 + R_2/R_1) + (QCC/A)R_2$$

The LM340-5.0 data sheet gives a maximum load regulation Lr = 50 mV and ΔI_Q = 1 mA for a 1-A load change.

$$Z_{340} = 50\text{ mV}/1\text{ A} = 0.05\text{ ohm}$$

$$QCC/A = 1\text{ mA}/1\text{ A} = 100\ \mu A/A$$

Using these values, the worst-case DC output impedance (AC output impedance is a function of C_2) is

$$Z_{OUT} = 0.05\ (1 + 310/300) = (100\ \mu A/A)\ 310$$

$$Z_{OUT} = 0.10\text{ ohm} + 0.03\text{ ohm} = 0.13\text{ ohm}$$

or a worst-case change of approximately 1.5% for a 1-A load change. Typical measured values are about one-third of the worst-case value.

7.16.4 Variable-Output Regulator

Figure 7–80 shows the LM340 connected as a regulator with variable output voltage. In this circuit, the ground terminal of the regulator is "lifted" by an amount equal to the voltage applied to the noninverting input of the op-amp LM101A. The output voltage of the regulator is therefore raised to a level set by the value of the resistive divider R_1, R_2, R_3 and limited by the input voltage. With the resistor values shown, the output voltage is variable from 7 to 23 V, and the maximum output current (pulsed load) varies from 1.2 to 2 A with a junction temperature of 25°C.

Because the LM101A is operated with a single supply (the negative supply pin is grounded), the common-mode voltage V_B must be at least a 2.0 $V_{BE} + V_{SAT}$ above ground. Also, the bias current I_B of the op-amp should be negligible compared to the current flowing through the resistive divider.

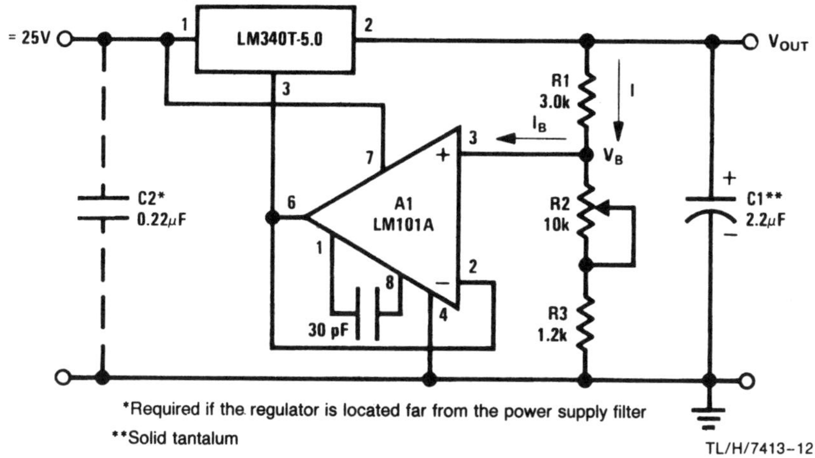

Figure 7–80. Variable-output regulator (National Semiconductor, *Linear Applications Handbook,* 1991, p. 347)

7.16.5 Variable-Output Regulator with Low Minimum Input

Figure 7–81 shows the LM340 connected as a variable-output regulator where the minimum output is below the nominal voltage (0.5 to 29 V). This circuit requires a separate negative supply. The voltage V_G at the ground pin of the regulator is de-

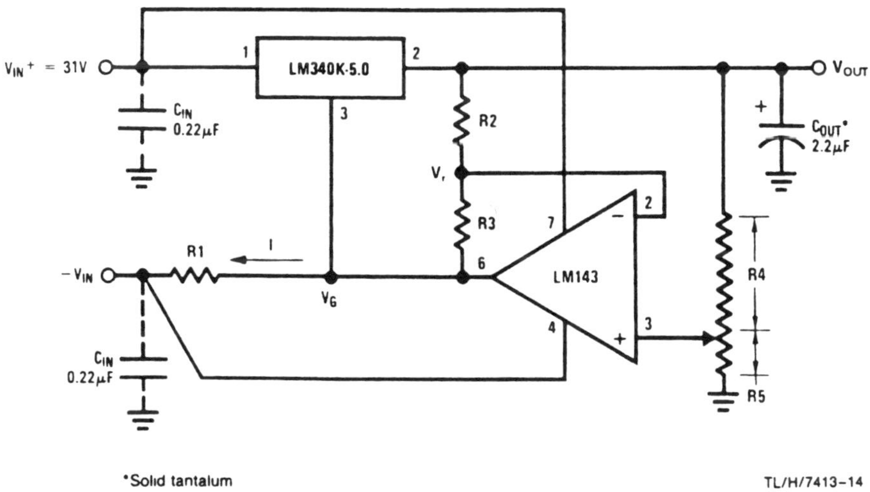

Figure 7–81. Variable-output regulator with low minimum input (National Semiconductor, *Linear Applications Handbook,* 1991, p. 348)

termined by the drop across R_1 and the amplifier gain. The current I may be determined by the relationship

$$I = \frac{V_{340}}{R_1} \frac{(R_2R_5) - (R_3R_4)}{R_4(R_2 + R_3)} + \frac{V_{IN}}{R_1}$$

or, if $R_2 + R_3 = R_4 + R_5 = R$,

$$I = \frac{V_{340} R_2}{R_1R_4} + \frac{1}{R_1} (V_{IN-} - V_{340})$$

Considering that the output is given by

$$V_{OUT} = V_G + V_{340}$$

and

$$V_G = R_1 I - V_{IN^-}$$

the equations can be combined to find output voltage as

$$V_{OUT} = V_{340} \frac{R_2}{R_4}$$

Notice that the output voltage is inversely proportional to R_4, so the output voltage may be adjusted very accurately for low values. A minimum output of 0.5 V has been set. This implies that

$$\frac{R_2}{R_4} = 0.1 \quad \frac{R_3}{R_4} = 0.9 \quad \frac{R_3}{R_2} = 9$$

An absolute zero output voltage would require that R_4 be infinity or R_2 be zero, neither of which is practical. The maximum output voltage is 30 V if the high-voltage op-amp LM143 is used. If only low values of V_{OUT} are needed, then an LM101 may be used. R_1 can be computed from

$$R_1 = \frac{V_{IN^-}}{IQ_{340}}$$

Figure 7–82 shows the load regulation as a function of output voltage.

7.16.6 Dual Linear Supply

Figure 7–83 shows an LM340 and an LM320 connected to form a dual linear supply. In this configuration, the regulators show both line and load regulation consistent with specifications. Operation is entirely normal until the problem of common loads occurs. For example, a 30-ohm load from the +15-V output to the −15-V output (representing a 0.5-A starting load for the LM340K-15, if the LM320K-15 is already started) will allow start-up in most cases.

The germanium power "diode" D_1 (made from a 2N3612 PNP) is included to ensure LM340 start-up over the full temperature range (into a worst-case 1-A cur-

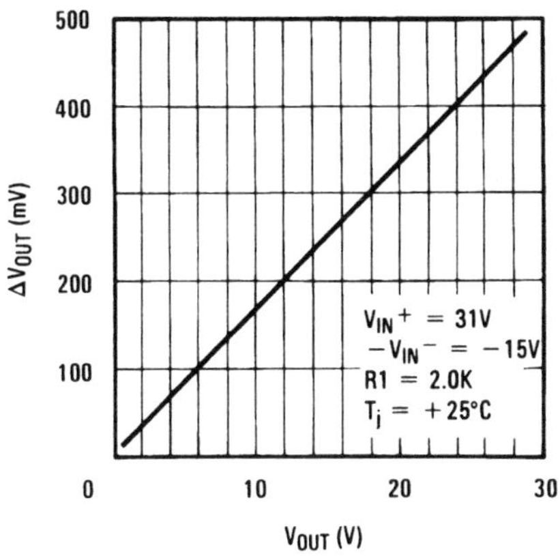

TL/H/7413–15

Figure 7–82. Load regulation as a fraction of output voltage (National Semiconductor, *Linear Applications Handbook,* 1991, p. 349)

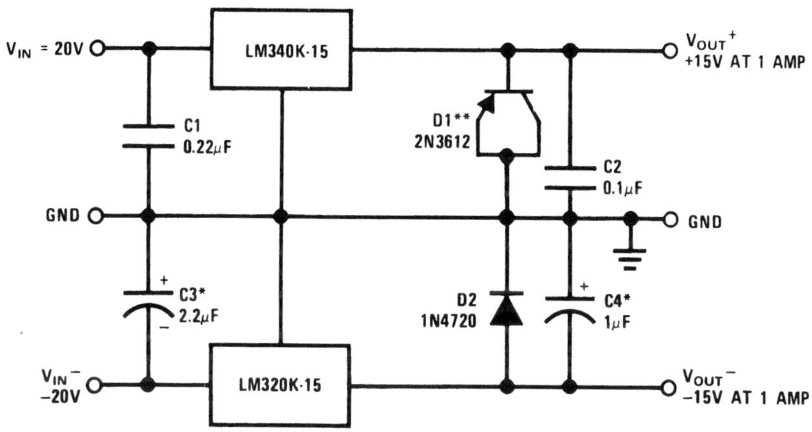

*Solid tantalum

**Germanium diode (using a PNP germanium transistor with the collector shorted to the emitter)

Note: C1 and C2 required if regulators are located far from power supply filter.

TL/H/7413–16

Figure 7–83. Dual linear supply (National Semiconductor, *Linear Applications Handbook,* 1991, p. 349)

rent-sink load). Because the forward voltage drop of the germanium D_1 is less than that of the silicon-substrate diode of the LM340, D_1 will take any fault current and allow the LM340 to start up, even into a negative-voltage load.

Diodes D_1 and D_2 also protect the regulator outputs from inadvertent shorts between outputs and to ground. For shorts between outputs, the voltage difference between either input and the opposite regulator output should not exceed the maximum rating of the device.

The circuit of Fig. 7–83 is a symmetrical +15-V supply for linear circuits. The same principle applies to nonsymmetrical supplies such as +5-V and –12-V regulators (say, as a supply for registers).

7.16.7 Tracking Dual Linear Supplies

Figure 7–84 shows an LM340 and an LM320 connected to form a tracking dual linear supply (±5 V to ±18 V). In this circuit, a fraction of the negative output voltage "lifts" the ground pins of the negative LM320K-15 regulator and the LM340K-15, through a voltage-follower and an inverter, respectively. The dual op-amp LM1588 permits the regulator output to be set between 5 and 20 V (because the LM1588 supply may go as high as ±22 V).

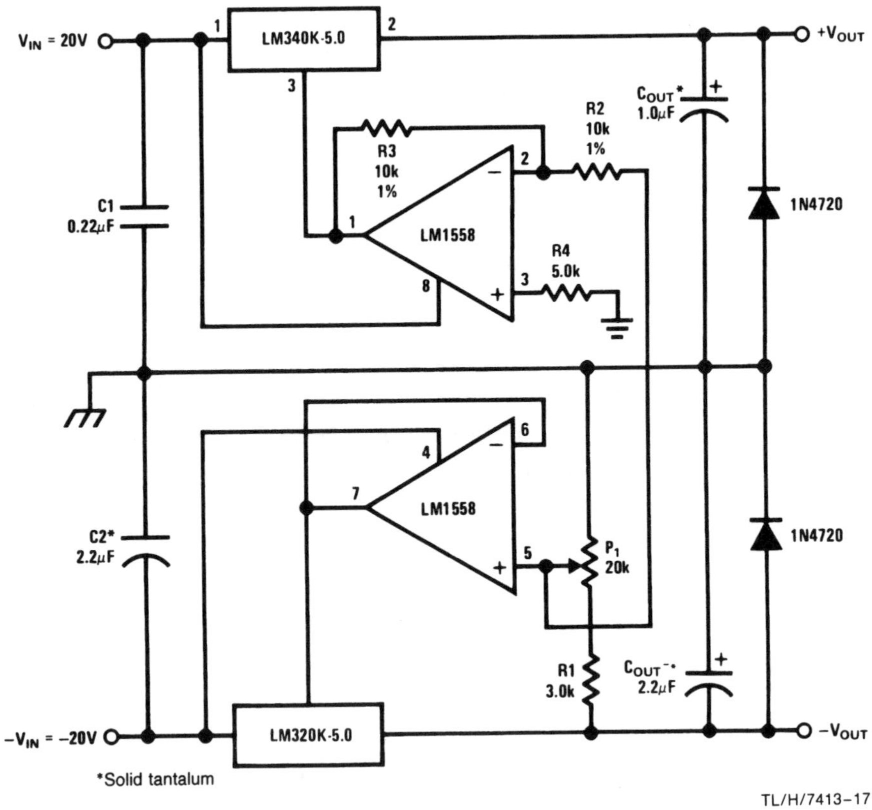

Figure 7–84. Tracking dual linear supply (National Semiconductor, *Linear Applications Handbook,* 1991, p. 350)

Because of the tighter output tolerance and the better drift of the LM320, the positive regulator is made to track the negative. The best tracking action is achieved by matching the gain of both op-amps. That is, resistors R_2 and R_3 must be matched as closely as possible.

If R_2 and R_3 are matched to better than 1%, the LM340 tracks the LM320 within 40–50 mV over the entire output range. Typical load regulation at a V_{OUT} of ±15 V for the LM340 is 40 mV from a 0- to 1-A pulse load, and 80 mV for the LM320.

Figure 7–85 shows an LM340 and an LM320 connected to form a symmetrical dual supply. Again, the positive regulator tracks the negative. Under steady-state conditions, V_A is at a virtual ground and V_B is at a V_{BE} above ground. Transistor Q_2 then conducts the quiescent current of the LM340.

If V_{OUT} becomes more negative, the collector-base junction of Q_1 is forward-biased, lowering V_B and raising the collector voltage of Q_2. As a result, $+V_{OUT}$ rises and the voltage V_A again reaches ground potential.

Assuming that Q_1 and Q_2 are perfectly matched, the tracking action remains unchanged over the full operating temperature range. With R_1 and R_2 matched to 1%, the positive regulator tracks the negative within 100 mV (less than 1%). Capacitor C_4 is added to improve stability. Typical load regulation for the positive and negative circuits (with a 0- to 1-A pulsed load; t_{ON} = 1 ms and t_{OFF} = 200 ms) are 10 mV and 45 mV, respectively.

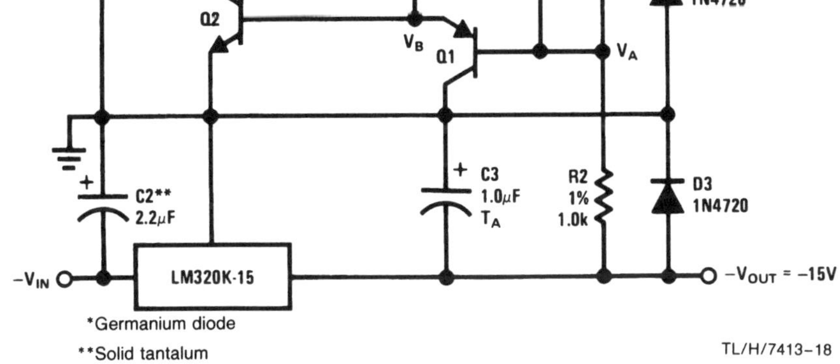

Figure 7–85. Tracking dual linear supply ±15 V (National Semiconductor, *Linear Applications Handbook,* 1991, p. 350)

7.16.8 Operating with High-Input Voltages

Figures 7–86 and 7–87 show how the LM340 can be operated with input voltages far in excess of the rated value. Typically, the input voltage of the LM340 must be kept within the limits specified in the data sheet. (This is true of any three-terminal regulator!) If the device is operated above the absolute maximum input-voltage rating, two failure modes might occur.

First, with the output shorted to ground, the series-pass transistor within the regulator will go into avalanche breakdown. Second, even if the output is not grounded, the feedback transistor within the regulator might fail. Typically, the feedback transistor is operated a few volts from the input. An excessive input results in an excessive collector-emitter voltage.

If the only available supply runs at a voltage higher than the maximum specified for the three-terminal regulator, one of the simplest ways to protect the regulator is to connect a zener in series or parallel with the input. This level-shifts the input voltage to a safe value. Although simple, this technique has an obvious drawback. The zener must dissipate a power equal to the supply voltage (minus the maximum input-voltage rating of the regulator) times the maximum output current. This value can easily be several watts—well beyond the capability of most zeners.

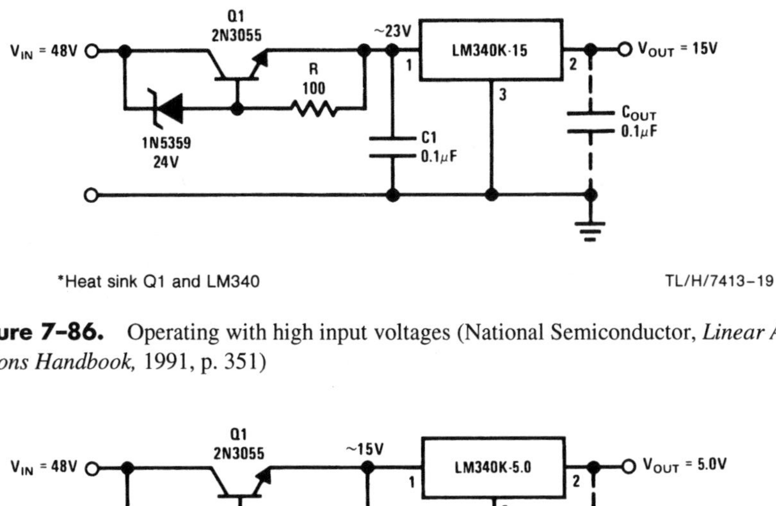

Figure 7–86. Operating with high input voltages (National Semiconductor, *Linear Applications Handbook,* 1991, p. 351)

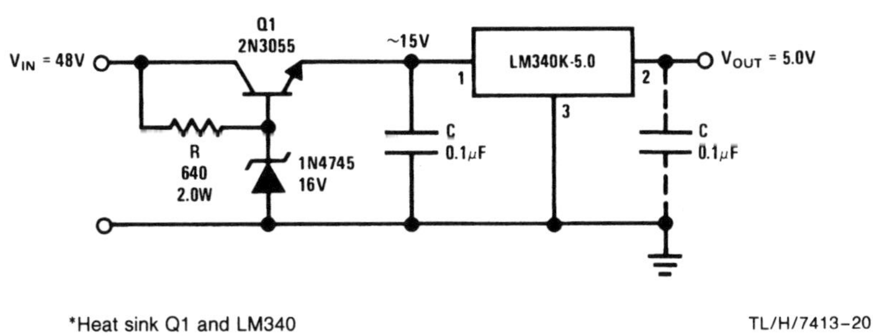

Figure 7–87. Operating with high input voltages (alternate) (National Semiconductor, *Linear Applications Handbook,* 1991, p. 351)

Figures 7–86 and 7–87 show more practical ways to overcome the power-dissipation problem. These circuits use inexpensive NPN/zener combinations as an equivalent to the power zener. Typical load regulation of the circuit in Fig. 7–86 is 40 mV, with a 0- to 1-A pulsed load (junction temperature 25°C). Line regulation is 2 mV for 1-V variation of the input voltage (with an I_{OUT} of zero). With the optional output capacitor, the measured noise is 700 μV peak-to-peak.

7.16.9 Operating with High Input and Output Voltages

Figure 7–88 shows how the LM340 can be operated with both input and output voltages in excess of the rated value. In this circuit, the output voltage is raised using a zener connected to the ground pin. Such a configuration increases the output level to that of the regulator, plus the breakdown voltage of the zener. Because the input voltage of the regulator is allowed to go as high as 80 V, a level-shifting NPN/zener is added to keep the voltage across the LM340 under permissible values.

The disadvantages of this circuit are increased output noise and possible output-voltage drift, because of the added diodes. Going from no-load to full-load conditions, the degraded regulation caused by D_1 is approximately V_Z (at 35 mA + I_Q) – V_Z (at I_Q). This formula translates into a load regulation of 60 mV for a variation in output current from 5 mA to 1 A (pulsed load) and a line regulation of 0.01% of input-voltage change (with I_{OUT} at 500 mA). Typical noise is 2 mV peak-to-peak (with C_2 = 0.1 μF).

To calculate the approximate value of R_1, subtract the combined zener voltages (36 + 33 = 69 V) from the input voltage (80 V), divide the result (11 V) by the

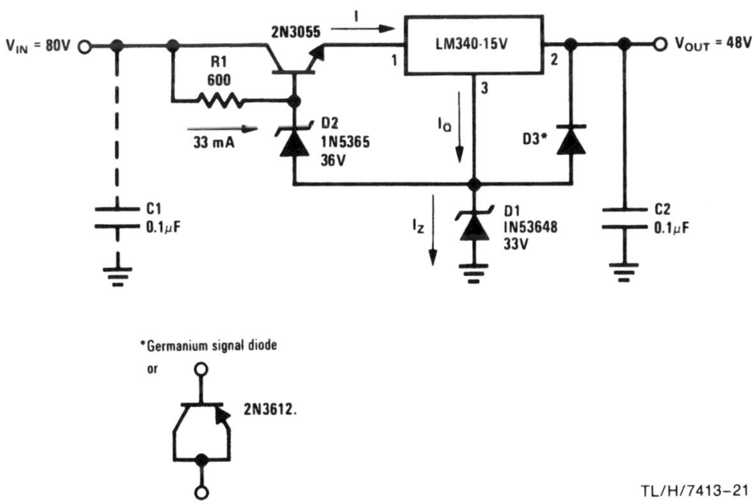

Figure 7–88. Operating with high input and output voltages (National Semiconductor, *Linear Applications Handbook,* 1991, p. 351)

desired full-load output current (1 A), and then multiply by the beta (55) of the 2N3055 transistor.

7.16.10 Electronic Shutdown

Figure 7–89 shows a practical method of shutting down the LM340 under control of a TTL or DTL logic gate. The series-pass transistor Q_1 operates either as a saturated transistor or as an open switch. With the logic input high (2.4 V specified minimum for TTL), transistor Q_2 turns on and pulls 50 mA down through R_2. This provides sufficient base drive to maintain Q_1 in saturation during the on-condition of the switch. When the logic is low (0.4 V specified maximum for TTL), Q_2 is held off, as is Q_1, and the switch is in the off-condition.

Typical turn-on time is 7 μs for resistive loads from 15 ohms to infinity. Turnoff time varies from about 3 μs for a 15-ohm load to 3 ms for a no-load condition. Turnoff time is controlled primarily by the time constant of R_{LOAD} and C_1.

7.16.11 Variable-Output Regulator with Overvoltage Shutdown

Figure 7–90 shows a high-voltage, variable-output regulator using the LM340 as the basic control element. One problem with this circuit is that the voltage across the regulator must be limited to maximum rating of the device. (The higher the input voltage, the higher the ground pin must be lifted.) As a result, the range of the vari-

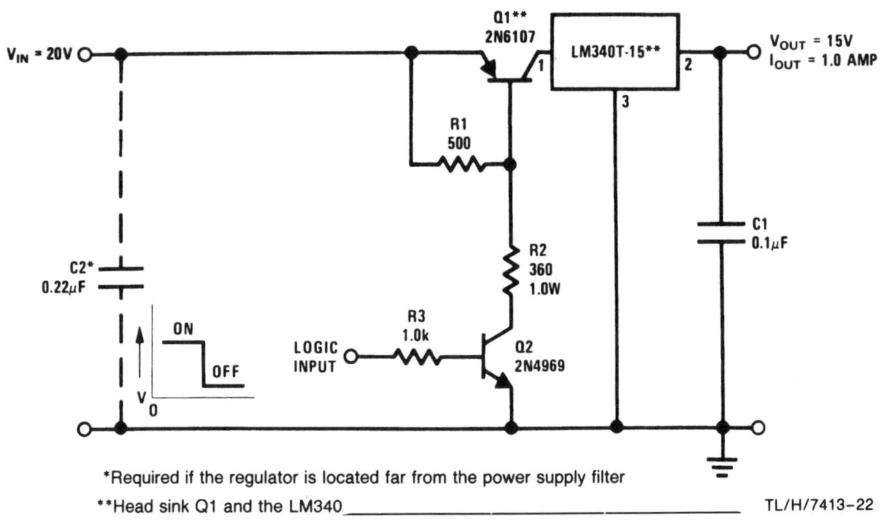

Figure 7–89. Electronic shutdown (National Semiconductor, *Linear Applications Handbook,* 1991, p. 352)

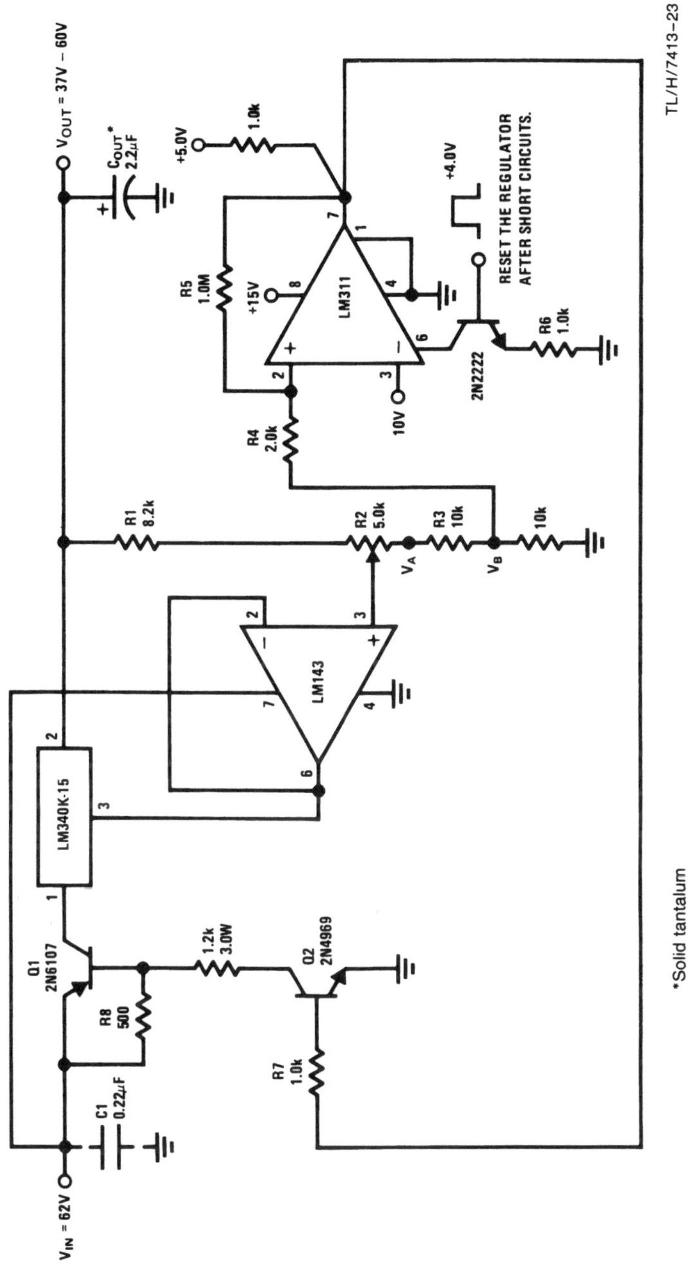

Figure 7–90. Variable-output regulator with overvoltage shutdown (National Semiconductor, *Linear Applications Handbook*, 1991, p. 352)

able output is limited by the supply-voltage limit of the op-amp and the maximum voltage allowed across the regulator. This can be estimated by

$$V_{OUT\ MAX} - V_{OUT\ MIN} = V_{SUPPLY\ MAX340} - V_{NOMINAL340} - 2\ V.$$

For example, using an LM340-15, $V_{OUT\ MAX} - V_{OUT\ MIN} = 35 - 15 - 2 = 18$ V.

Even though the LM340 has short-circuit protection, when the output drops, V_A also drops and the voltage difference across the device increases. If the differential exceeds 35 V, the series-pass transistor within the LM340 will break down.

The circuit includes an overvoltage-shutdown provision to prevent such breakdown. When the output drops, the comparator switches low. This pulls the base of Q_2 down, thus opening switch Q_1. With Q_1 open, the LM340 is shut down.

Once the short circuit is removed, the LM311 must be activated through the strobe (+4-V pulse) at pin 6 (through the 2N2222 transistor). This closes Q_1 and starts the regulator action. The additional voltages required to operate the comparator may be taken from the 62-V input.

7.17 Precision Dual-Tracking Regulators

This section describes a number of circuits based on the use of dual-tracking regulators (the classic National Semiconductor LM125/LM126). These devices are precision IC voltage regulators. Each provides separate positive and negative regulated outputs, thus simplifying dual linear-supply designs. Operation requires few or no external components, depending on the application. Internal settings provide fixed output voltages: the LM125 at ±15 V and the LM126 at ±12 V.

Each regulator is protected from excessive internal power dissipation by the usual thermal-shutdown circuit, which turns off the regulator whenever the chip reaches a preset maximum temperature (a T_J of about 175°C). Other features include both internal and external current-limit sensing for device protection while operating with or without external current boost.

For applications requiring more current than the internal current limit will allow, boosted operation is possible with the addition of one NPN series-pass transistor per regulator. External resistors sense load current for controlling the limiting circuits. Internal frequency compensation is provided on both positive and negative regulators. The internal voltage-reference pin is brought out to facilitate noise filtering when desired.

7.17.1 Internal Circuit Description

Figure 7–91 shows a block diagram of the basic dual-tracking regulator. A voltage reference establishes a fixed DC level, independent of supply or temperature variations, at the noninverting input to the negative error amplifier. This amplifier drives the negative output-control circuit, which includes the high-current output transistors and current-limiting and thermal-shutdown circuits.

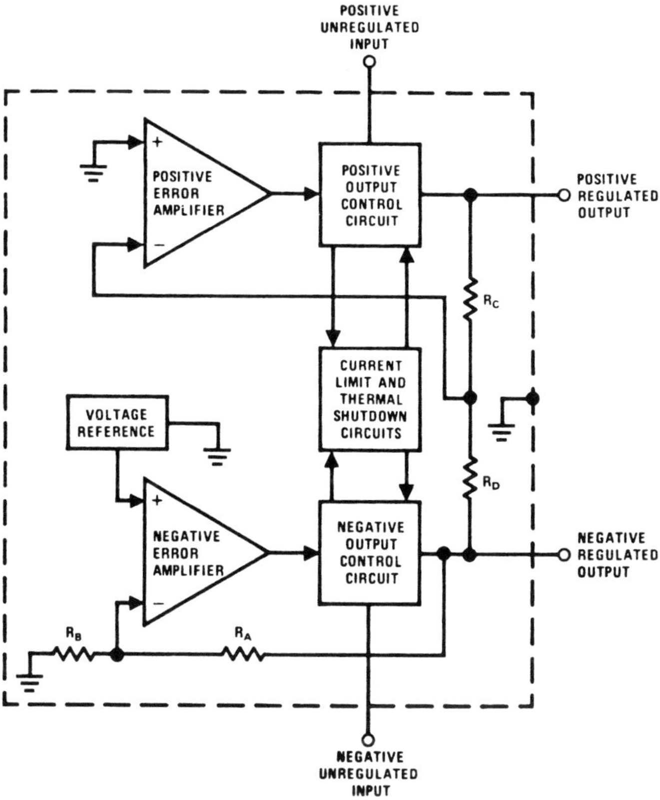

Figure 7–91. Block diagram for LM125/LM126 dual-tracking regulator (National Semiconductor, *Linear Applications Handbook,* 1991, p. 305)

The negative-regulator output voltage is established by comparing the voltage reference against a fraction of the output, as set by R_A and R_B. To get the desired tracking action of the positive regulator, a voltage established between the positive and negative regulator outputs by resistors R_C and R_D is compared to ground by the positive error amplifier. This ensures that the positive-regulator output voltage will always equal the negative-regulator output voltage, multiplied by the ratio of R_C to R_D.

The positive output-control circuit is essentially the same as that in the negative regulator. The current-limit and thermal-shutdown circuits sense the output load current and die temperature, respectively, and switch off all output-drive capability upon reaching the determined limits.

7.17.2 Basic Dual-Tracking Regulator

Figure 7–92 shows the basic regulator configuration with no external-boost transistors. The only connections required (other than inputs, outputs, and ground)

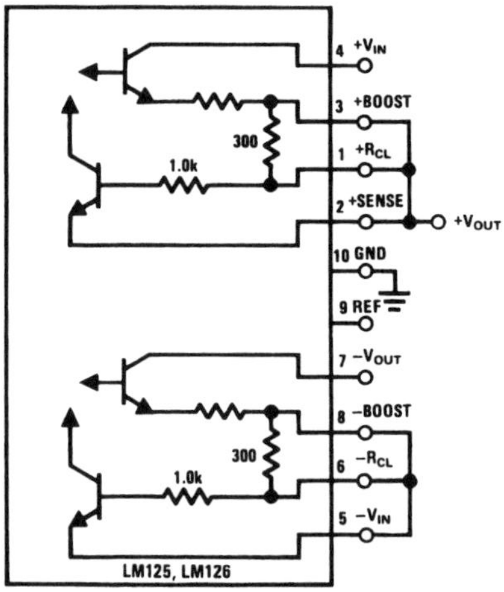

Note: Pin numbers for metal can package only. TL/H/7390–8

Figure 7–92. Basic dual-tracking regulator (National Semiconductor, *Linear Applications Handbook,* 1991, p. 309)

are to complete the output-current paths from $+R_{CL}$ to $+V_{OUT}$ and from $-R_{CL}$ to $-V_{IN}$.

These connections may be a direct short if the internal preset current-limit is desired, or resistors may be used to set the maximum current at some level less than the internal current limit. The internal 300-ohm resistors from pins 3 to 1 and pins 8 to 6 should be shorted as shown when no external series-pass transistors are used.

To improve ripple rejection and transient response, filter capacitors may be added to the inputs, outputs, or both, depending on the unregulated input available. If a very low noise output voltage is desired, a capacitor may be connected from the reference voltage pin (9) to ground. This shunts noise generated by the internal reference zener. Figure 7–93 shows the internal current-limiting characteristics for the basic regulator circuit of Fig. 7–92.

7.17.3 Boosted-Current Regulator

Figure 7–94 shows the connections for boosted-current operation with external series-pass transistors. As shown, an NPN series-pass transistor is added to each regulator. This increases the maximum output current by a factor of the external transistor beta.

In the boosted configuration, current limiting is usually necessary to ensure that the external series-pass device is not overheated or destroyed. If the regulator output is grounded, the series-pass device may fail and short, destroying the regulator.

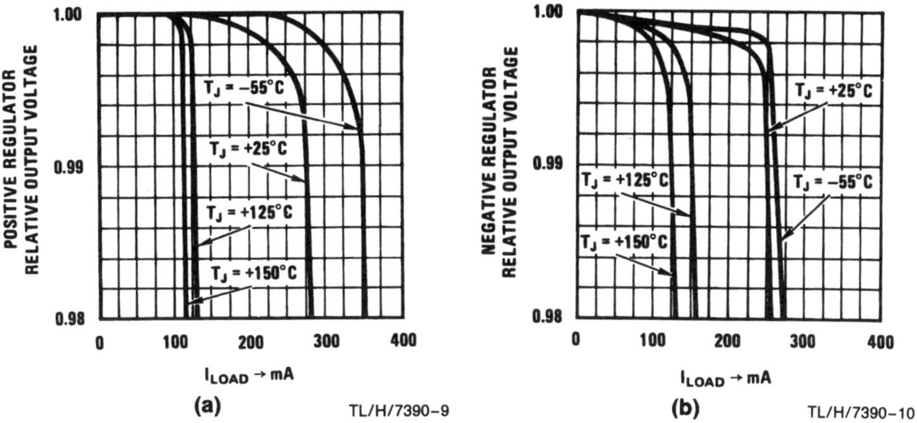

Figure 7–93. Internal current-limiting characteristics of LM125/LM126 (National Semiconductor, *Linear Applications Handbook,* 1991, p. 309)

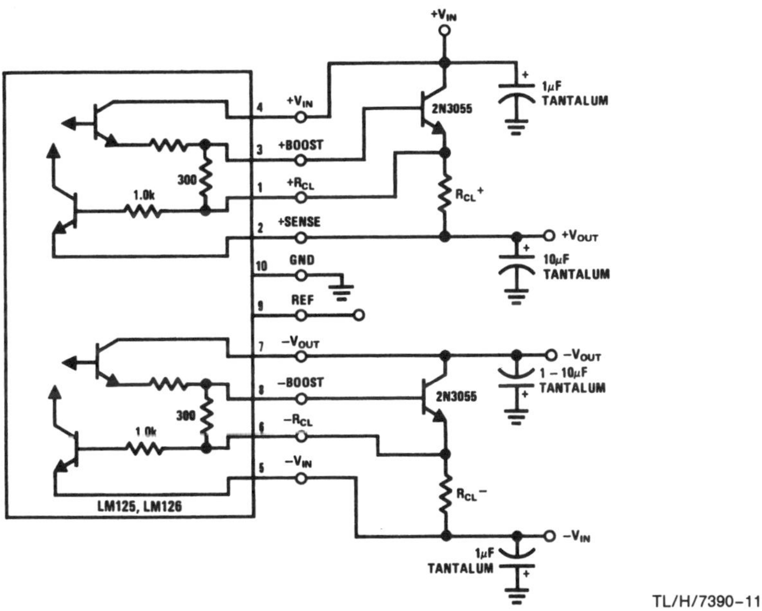

Figure 7–94. Boosted-current dual-tracking regulator (National Semiconductor, *Linear Applications Handbook,* 1991, p. 309)

Series resistor R_{CL} is used to sense load current. (R_{CL} is the equivalent of R_{SC} in Figs. 4-2 and 4-3.) The regulator will current-limit when the voltage drop across R_{CL} equals the current-limit sense voltage found in Fig. 7–95. Figure 7–96 shows

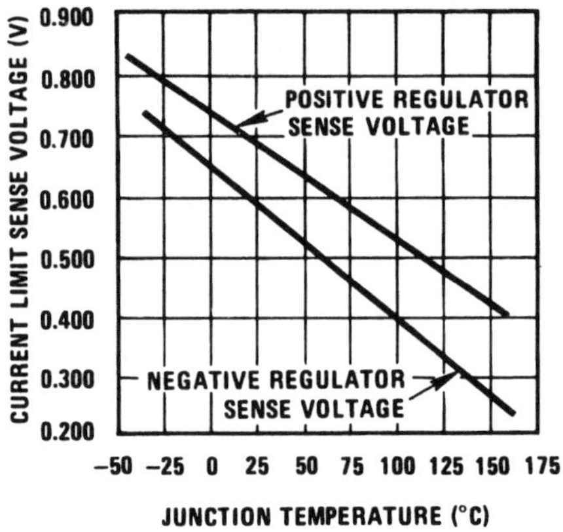

Figure 7–95. Current-limit sense voltage versus junction temperature (National Semiconductor, *Linear Applications Handbook,* 1991, p. 310)

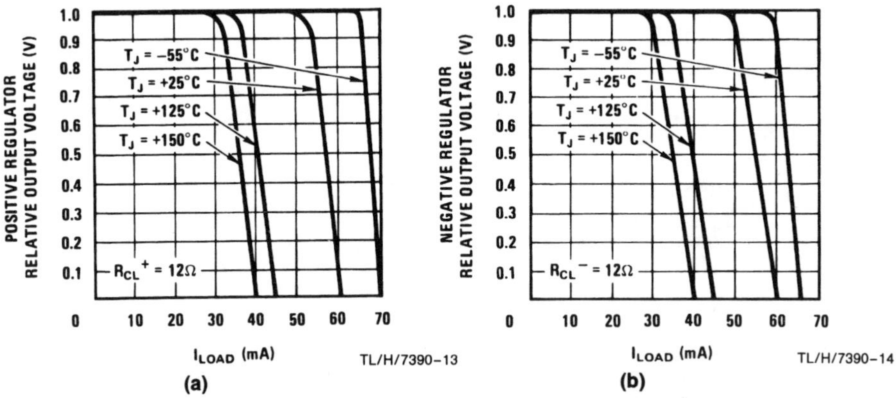

Figure 7–96. External current-limit characteristics (unboosted) (National Semiconductor, *Linear Applications Handbook,* 1991, p. 310)

the external current-limit characteristics unboosted, while Fig. 7–97 shows the external current-limit characteristics in the boosted mode.

Notice that the current-limit sense voltage (Fig. 7–95) is 0.6 to 0.7 V at a junction temperature of 25°C. This is typical for the many regulators described in this chapter, and in Chapter 4. For example, if shutoff is desired at 1 A (in the boosted configuration), use a value of 0.6 ohm (0.6 V/1 A = 0.6 ohm). This is shown in Fig. 7–97.

To ensure circuit stability at high currents in the boosted configuration of Fig. 7–94, it may be necessary to bypass each input with low-inductance tantalum capacitors to prevent forming resonant circuits with long input leads. A capacitance of at

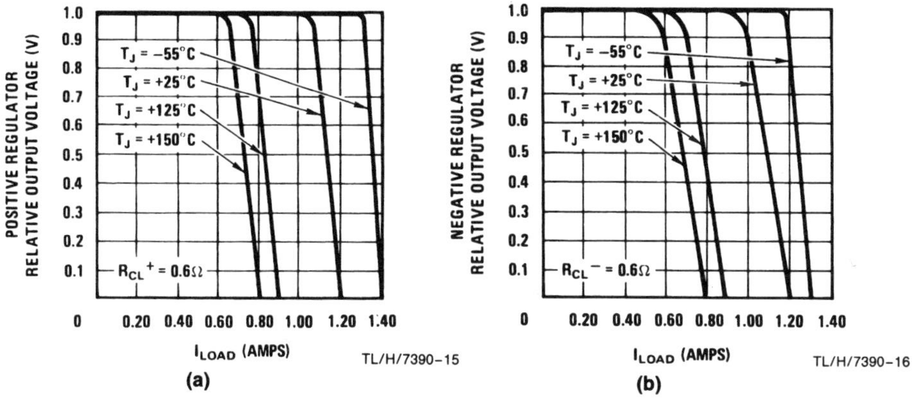

Figure 7–97. External current-limit characteristics (boosted) (National Semiconductor, *Linear Applications Handbook,* 1991, p. 310)

least 1 μF is recommended at the input. The same problem can also occur at the regulator output where a 10-μF or larger tantalum capacitor will ensure stability and increase ripple rejection.

The 2N3055 series-pass transistor is low in cost and maintains a reasonably high beta at collector currents up to several amps. The 2N3055 may be of either planar or alloy junction construction. The planar devices have a high f_T (frequency at which beta = 1) and thus provide more stable operation because of low phase shift. The alloy devices, with f_T typically less than 1 MHz, may require additional compensation to guarantee stability.

The simplest compensation for the alloy devices is to use output filter-capacitor values greater than 50 μF (tantalum). An alternative is to use an RC filter to create a leading phase response to cancel some of the phase lag. The stability problem with slower series-pass transistors, if it occurs at all, is usually seen only on the negative regulator. This is because the positive-regulator output stage is a conventional Darlington, while the negative output stage contains three devices in a modified triple-Darlington configuration, giving slightly more internal phase shift.

Additional compensation may be added to the negative regulator by connecting a small capacitor (in the 100-pF range) from the negative boost terminal to the internal reference. Because the positive regulator uses the negative-regulator output for a reference, this also offers some additional indirect compensation for the positive regulator.

7.17.4 Foldback Current Limiting

Figure 7–98 shows the LM125 connected in the foldback current-limit configuration. (This is the same as described in Section 4.6 and shown in Fig. 4-9.)

In many regulator applications, the normal operating power dissipation in the series-pass transistor can easily be multiplied by a factor of 10 or more when the

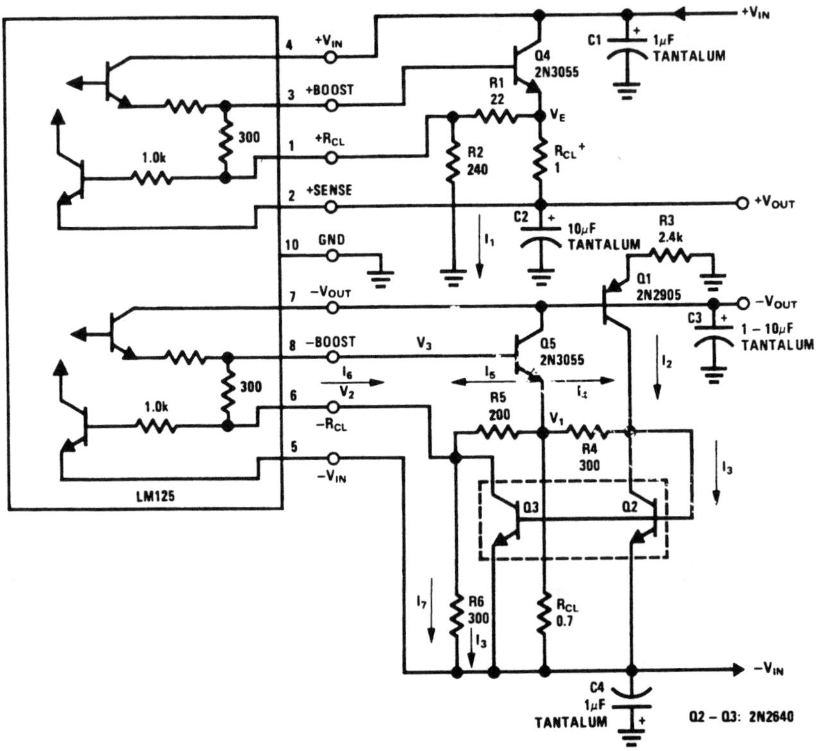

Figure 7–98. Dual-tracking regulator with foldback current-limit (National Semiconductor, *Linear Applications Handbook,* 1991, p. 312)

output is shorted. This might destroy the series-pass device, and possibly the regulator, unless a heat sink is oversized to handle the fault condition. A foldback current-limiting circuit reduces short-circuit output current to a fraction of the full-load output current, thus avoiding the need for the larger heat sink.

In foldback current limiting, a fraction of the output voltage must be used to oppose the voltage across the current-limit sense resistor (R_{CL}). Current limiting does not occur until the voltage across the sense resistor is higher than this opposing voltage by the amount shown in Fig. 7–95.

When the output is grounded, the opposing voltage is no longer present, so current limiting occurs at a lower level. This is accomplished in Fig. 7–98 by a programmable current source to give a constant voltage drop across R_5 for the negative regulator, and by a simple resistor divider for the positive regulator. The reason for the difference between the two is that the negative-regulator current-limiting circuit is located between the output pass transistor and the unregulated input, while the positive-regulator current limiter is between the output pass transistor and the regulated output.

In the positive regulator, a voltage divider (R_1 and R_2 from V_E to ground) creates a fixed voltage drop across R_1 opposite in polarity to the drop across $R_{CL}+$.

When the load current increases to the point where the drop across $R_{CL}+$ is equal to the drop across R_1, plus the current-limit sense voltage given in Fig. 7–95, the positive output will begin to drop.

As positive output begins to drop, the voltage across R_1 also decreases so that less load current is required to produce the current-limit sense voltage. With the regulator output fully shorted to ground ($+V_{OUT} = 0$), the current limit is set by the value of R_{CL} alone.

If the feedback current (I_{FB}) divided by the short-circuit current (I_{SC}) is less than 5, then the following equations can be used to calculate the positive-regulator foldback current-limiting resistor values:

$$R_{CL}+ = \frac{V_{SENSE}}{I_{SC}}$$

where V_{SENSE} is from Fig. 7–95 and I_{SC} is the short-circuit current.

At the maximum load-current foldback point:

$$V_{R_{CL+}} = I_{FB}\, R_{CL}+$$

$$V_{R_1} = V_{R_{CL+}} - V_{SENSE}$$

$$V_{R_1} = I_{FB}\, R_{CL}+ - V_{SENSE}$$

Then

$$R_1 = \frac{V_{R_1}}{I_1}$$

and

$$R_2 = \frac{+V_{OUT} + V_{SENSE}}{I_1}$$

The only point of caution is to ensure that the total current (I_1 shown in Fig. 7–98) through R_2 is much greater than the current contribution from the internal 300-ohm resistor. This can be checked by the expression

$$\frac{I_{FB}\, R_{CL}+}{300} << I_1$$

Now let us work out a design example using Fig. 7–98 and the following characteristics:

$I_{FOLDBACK} = 2$ A

$I_{SC} = 500$ mA

V_{SENSE} (See Fig. 7–95)

$+V_{IN} = 25$ V

$+V_{OUT} = 15$ V

Beta of the series-pass transistor = 70

$\theta_{JA} = 150°C/W$

$T_A = 50°C$

With a beta of 70 in the series-pass device and a maximum output current of 2 A, the regulator must deliver

$$\frac{2A}{beta} = \frac{2A}{70} = 29 \text{ mA}$$

The LM125 power dissipation will be calculated ignoring any negative output current for this example:

$$P_{LM125} = (V_{IN} - V_{OUT})\, I_{OUT}$$
$$= (25 - 15)\, 29 \text{ mA}$$
$$= 290 \text{ mW}$$

$$T_{RISE} @ \theta_{JA} = 150°C/W = 150°C \times 0.29 \text{ W} = 44°C$$

$$T_J = T_A + T_{RISE} = 50°C + 44°C = 94°C$$

From Fig. 7–95:

$$V_{SENSE} @ (T_J = 94°C) = 520 \text{ mV}$$

Then

$$R_{CL^+} = \frac{V_{SENSE}}{I_{SC}} = \frac{520 \text{ mV}}{500 \text{ ma}} \approx 1 \text{ ohm}$$

and

$$V_{R_{CL^+}} = I_{FB}\, R_{CL^+} = (2 \text{ A})\,(1 \text{ ohm}) = 2 \text{ V}$$

and

$$V_{R_1} = V_{R_{CL^+}} - V_{SENSE}$$
$$V_{R_1} = 2 \text{ V} - 520 \text{ mV} = 1.480 \text{ V}$$

A value for I_1 can now be found:

$$\frac{I_{FB}\, R_{CL^+}}{300} = \frac{2 \text{ A} \times 1 \text{ ohm}}{300 \text{ ohms}} = 6.6 \text{ mA}$$

So let I_1 be 10 × 6.6 mA, or 66 mA:

$$R_1 = \frac{V_{R1}}{I_1} = \frac{1.480}{66 \text{ mA}} \approx 22 \text{ ohms}$$

$$R_2 = \frac{+V_{OUT} + V_{SENSE}}{I_1} = \frac{15 + 0.520}{66 \text{ mA}} \approx 240 \text{ ohms}$$

The foldback characteristics are shown in Fig. 7–99 for the positive-regulator circuit.

The negative-regulator foldback works essentially the same way. Transistor Q_1 forces a constant current I_2, determined by $-V_{OUT}$ and R_3, through Q_2. Transistors Q_2 and Q_3 are matched so a current identical to I_3 will flow through Q_3.

With the output short-circuited ($-V_{OUT} = 0$), Q_1 is off, setting I_2 at zero. The load current will be limited when V_1 increases sufficient (because of load current) to make V_2 higher than $-V_{IN}$ (by an amount equal to the current-limit sense voltage).

The short-circuit current is

$$I_{SC} \approx \frac{V_{SENSE}}{R_{CL-}}$$

To calculate the maximum full-load current with the output still in regulation, current I_2 is

$$I_2 = \frac{V_{OUT} - V_{BE_{Q1}}}{R_3}$$

where $V_{BE_{Q_1}}$ is the base-emitter voltage of Q_1. At the point of maximum load current I_{FB}, where the regulator should start folding back, voltages V_1 and V_2 are

$$V_1 = -V_{IN} + I_{FB} R_{CL-}$$

and

$$V_2 = -V_{IN} + V_{SENSE}$$

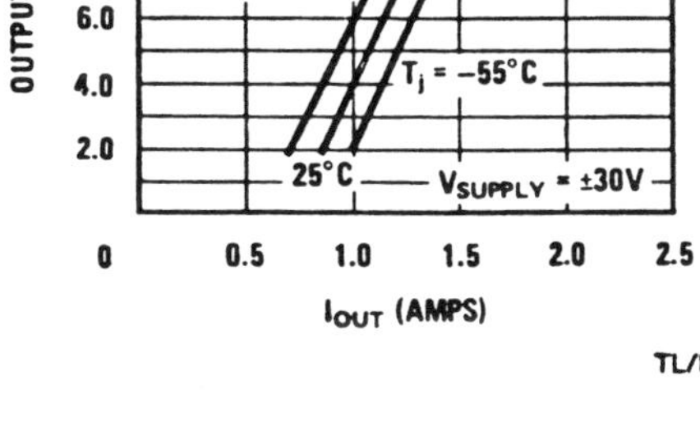

Figure 7–99. Foldback characteristics for dual-tracking regulator (National Semiconductor, *Linear Applications Handbook,* 1991, p. 313)

The current through Q_2 (and Q_3) will have increased from I_2 (by the amount of I_4) because of the voltage V_1 increasing above the no-load quiescent value. Because the voltage across Q_2 is simply the diode drop of a base-emitter junction:

$$I_4 = \frac{[V_1 - (-V_{IN})] - V_{BE}}{R_4}$$

By substituting the elements in the equation for V_1,

$$I_4 = \frac{(I_{FB}\,R_{CL^-}) - V_{BE}}{R_4} = \frac{(I_{FB}\,R_{CL^-}) - V_{BE}}{300}\ \text{ohms}$$

The current through Q_2 is

$$I_3 = I_2 + I_4$$

and the current through Q_3 is

$$I_3 = I_5 + I_6 - I_7$$

The drop across R_5 is found from

$$V_1 - V_2 = (-V_{IN} + I_{FB}\,R_{CL^-}) - [V_{SENSE} + (-V_{IN})]$$

This can be simplified to

$$V_1 - V_2 = (I_{FB}\,R_{CL^-}) - V_{SENSE}$$

Because V_{SENSE} is the base-to-emitter voltage drop of the internal limiter resistor, the V_{SENSE} in the simplified equation very nearly equals the V_{BE} in the substituted equation for I_4. Therefore, the drop across R_5 is approximately the same as the drop across R_4. The current through R_5 (I_5) can then be determined as

$$I_5 = \frac{(I_{FB}\,R_{CL^-}) - V_{SENSE}}{300}$$

Summing the currents through Q_3 is now possible, assuming that the base-emitter drop of the 2N3055 series-pass transistor can be given by $V_{BE} \approx V_{SENSE}$:

$$I_6 = \frac{V_3 - V_2}{300},$$

where

$$V_3 = V_1 + V_{BE} \approx V_1 + V_{SENSE}$$

$$I_6 = \frac{V_1 + V_{SENSE} - V_2}{300}$$

By substituting the elements in the equation for V_1,

$$I_6 = \frac{I_{FB}\,R_{CL^-}}{300}$$

and

$$I_7 = \frac{V_2 - (-V_{IN})}{R_6} = \frac{V_{SENSE}}{R_6}$$

Using the equations for current through Q_2 and Q_3, and inserting resistor values shown in Fig. 7–98, I_2 is found as follows:

$$I_2 + I_4 = I_5 + I_6 - I_7$$

$$I_2 + (I_{FB}\,R_{CL}-) - \frac{V_{SENSE}}{300} = I_5 + \frac{I_{FB}\,R_{CL}-}{300} - \frac{V_{SENSE}}{300}$$

By canceling, we find:

$$I_2 = I_5.$$

This is the key to the negative-foldback circuit. Current source Q_1 forces current I_2 to flow through resistor R_5. The voltage drop across R_5 opposes the normal current-limit sense voltage so that the regulator will not current-limit until the drop across $R_{CL}-$ (because of load current) equals the controlled drop across R_5 plus V_{SENSE} (given in Fig. 7–95). This can be written as

$$I_{FB} = \frac{V_{SENSE} + I_2\,R_5}{R_{CL}-}$$

$$I_{FB} = \frac{V_{SENSE} + 200\,I_2}{R_{CL}-}$$

Now let us work out a design example using Fig. 7–98 and the following characteristics:

$I_{FB} = 2.5$ A

$I_{SC} = 750$ mA

V_{SENSE} (see Fig. 7–95)

$-V_{IN} = 25$ V

$-V_{OUT} = -15$ V

Beta of the series-pass transistor = 90

$\theta_{JA} = 150°C/W$

$T_A = 25°C$

The same calculations are used here to figure V_{SENSE} as with the positive-foldback circuit. With a beta of 90 in the series-pass device and a maximum output current of 2.5 A, the regulator must deliver

$$\frac{2.5\text{ A}}{\text{beta}} = \frac{2.5\text{ A}}{90} = 28\text{ mA}$$

The LM125 power dissipation will again be calculated ignoring any negative output current for this example:

$$P_{LM125} = (V_{IN} - V_{OUT})\, I_{OUT}$$
$$= (25 - 15)\ 28\ \text{mA}$$
$$= 280\ \text{mW}$$

$$T_{RISE} @ \theta_{JA} = 150°\text{C/W} \times 0.28\ \text{W} = 42°\text{C}$$
$$T_J = T_A + T_{RISE} = 25°\text{C} + 42°\text{C} = 67°\text{C}$$

From Fig. 7–95:

$$V_{SENSE} @ T_J\ (= 67°\text{C}) = 500\ \text{mV}$$

Then

$$R_{CL^+} = \frac{V_{SENSE}}{I_{SC}} = \frac{500\ \text{mV}}{750\ \text{mA}} = 0.68\ \text{ohm}$$

$$I_2 = \frac{(I_{FB}\, R_{CL^-}) - V_{SENSE}}{R_5}$$
$$= \frac{(2.5\ \text{A})\ (0.68) - 0.5\ \text{A}}{200} = 6\ \text{mA}$$

$$R3 = \frac{V_{OUT} - V_{BE_{Q_1}}}{I_2}$$
$$= \frac{-15\ \text{V} - 0.7\text{V}}{6\ \text{mA}} = \frac{14.3\ \text{V}}{6\ \text{mA}} = 2.4\ \text{k}$$

Figures 7–99 and 7–100 show the measured foldback characteristics for the values derived in the design examples. The value of R_5 is set low so that the magnitude of I_5 for foldback is greater than I_4 through I_6. This reduces the foldback-point sensitivity to the temperature coefficient of the internal 300-ohm resistor, plus any mismatch in the temperature coefficient of Q_2, Q_3, or the series-pass transistor.

The value of R_6 can be computed as follows:

$$R_6 = \frac{V_{SENSE}}{I_7} = \frac{V_{SENSE}}{I_5 + I_6 + I_3}$$

and by combining, the value of R_6 is

$$R_6 = \frac{V_{SENSE^-}}{I_6 - I_4}$$

$$= \frac{V_{SENSE^-}}{I_{FB}\, R_{CL^-}\left(\frac{1}{300} - \frac{1}{R_4}\right) + \frac{V_{BE}}{R_4}}$$

By letting V_{BE} approximately equal V_{SENSE}, and $R_4 = 300$ ohms to match the internal 300 ohms, the value of R_6 is equal to R_4.

Also letting $I_4/I_5 = 2/3$, then $R_5 = 200$ ohms.

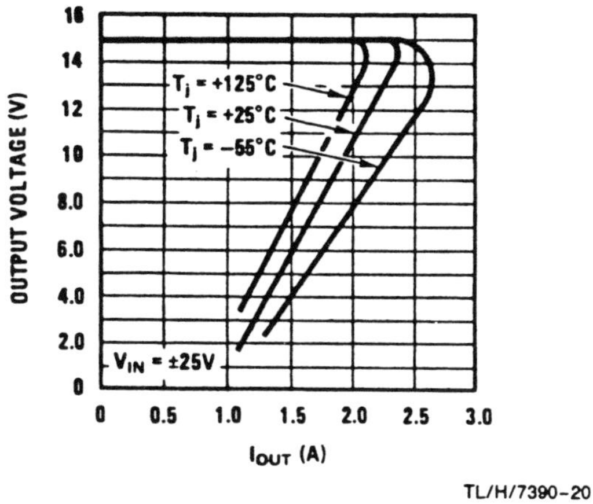

Figure 7–100. Foldback characteristics for negative dual-tracking regulator (National Semiconductor, *Linear Applications Handbook,* 1991, p. 314)

7.17.5 Electronic Shutdown

Figure 7–101 shows the LM125 connected for electronic shutdown of a boosted regulator. The electronic-shutdown characteristics are shown in Fig. 7–102.

In some regulated linear-supply applications, it is desirable to shut down the regulated outputs ($\pm V_O = 0$) without having to shut down the unregulated inputs (which may be powering additional equipment). Various shutdown methods may be used. The simplest is to insert a relay, a saturated bipolar device, or some other type of switch in series with either the regulator inputs or outputs. The switch must be able to open and close under maximum load current (which may be several amps).

As an alternate solution, the internal reference voltage of the regulator may be shorted to ground. This will force the positive and negative outputs to approximately +700 mV and +300 mV, respectively. Both outputs are fully active, so the full output current can still be supplied to a low-impedance load. If this is unacceptable in any particular application, another solution must be found.

The circuit of Fig. 7–101 provides complete electronic shutdown of both regulators. The shutdown control signal is TTL compatible, but by adjusting R_8 and R_9, the regulator may be shut down at any desired level above $2V_{BE}$, calculated as follows:

$$V_t \approx \left[\frac{R_8}{R_3 \text{ beta } Q_4} + \frac{R_9}{R_3}\right] V_{BE} + 2V_{BE}$$

Positive and negative shutdown operations are similar. When a shutdown signal V_T is applied, Q_4 draws current through R_3 and D_2, establishing a voltage V_R, which starts the current sources Q_1 and Q_2. Assuming that Q_1 and Q_2 are matched, and making $R_1 = R_2 = R_3$, the currents I_1, I_2, and I_3 are equal, and both sides of the regulator shut down simultaneously.

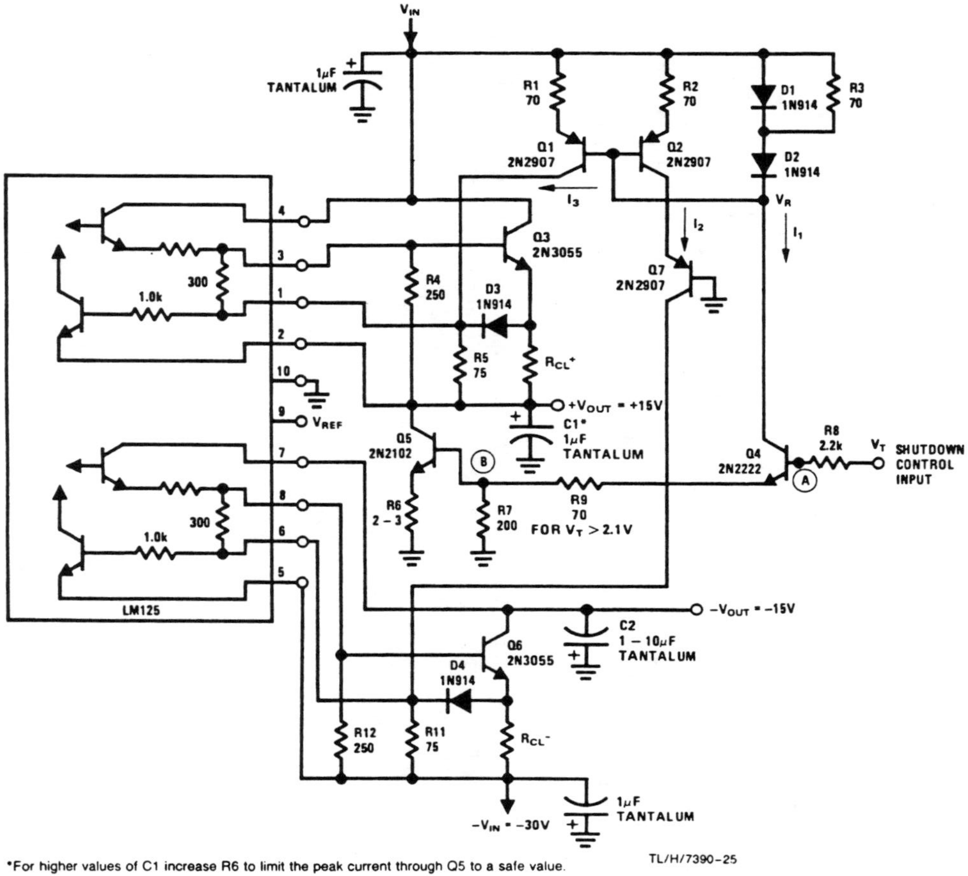

*For higher values of C1 increase R6 to limit the peak current through Q5 to a safe value.

Figure 7–101. Electronic shutdown for boosted regulator (National Semiconductor, *Linear Applications Handbook,* 1991, p. 318)

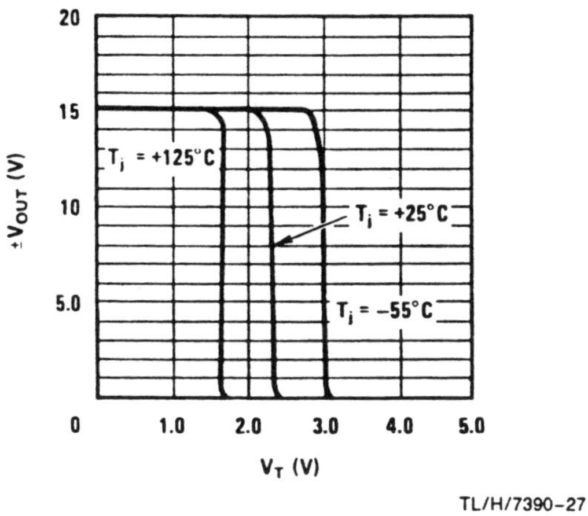

Figure 7–102. Electronic shutdown characteristics (National Semiconductor, *Linear Applications Handbook,* 1991, p. 319)

The current I_3 creates a drop across R_5, which equals or exceeds the limit-sense voltage of the positive regulator. This causes the positive regulator to shut down. Since I_3 has no path to ground except through the load, a fixed load is provided by Q_5, which is turned on by the variable current source Q_4.

Capacitor C_1 also discharges through Q_5 and current-limiting resistor R_6. Resistor R_4 prevents Q_3 from turning on during shutdown. (This could occur because of the drop across R_5, plus the internal 300-ohm resistor.) Diode D_3 prevents I_3 from being shunted through R_{CL}.

Capacitor C_2 discharges through the load. Transistor Q_7 shares the total supply voltage with Q_2, thus limiting power dissipation of Q_2.

Figure 7–103 shows an alternate circuit for the shutdown input. This alternate can be used when the design calls for a V_T of 2 V, and V_T is increased above the present threshold value. Under such circumstances, I_1 is increased and Q_4 must dissipate $(V_{IN} - 3V_{BE}) \times I_1$. The simplest solution is to increase the value of R_8. If this is not sufficient, a set of diodes may be added between nodes A and B to clamp I_1 to a reasonable value.

With the configuration shown in Fig. 7–103, I_1 is made independent of V_T. (If V_{BE} is assumed to be 0.7 V, then I_1 can be set to a reasonable value of 10 mA when R_9 is 70 ohms.) The regulator will shut down at any desired level above $3V_{BE}$, without overheating transistor Q_4. Notice that if the diodes are added as shown in Fig. 7–103, diode D_1 in Fig. 7–101 can be omitted.

7.17.6 Series-Pass Darlingtons

Figure 7–104 shows the LM125 connected for a boosted output using Darlington series-pass elements (2N3715 and 2N3772). With this configuration, the output

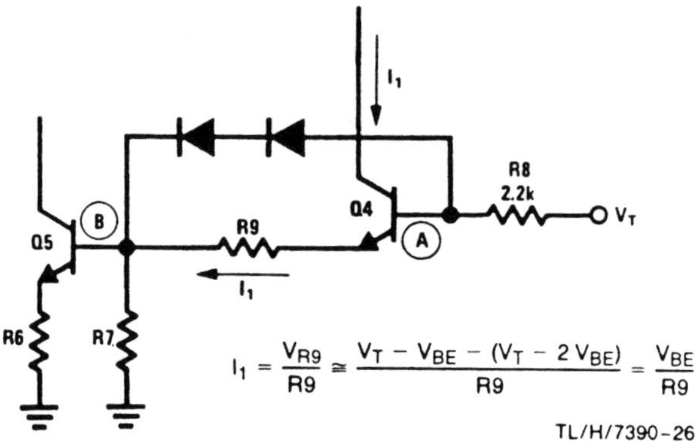

$$I_1 = \frac{V_{R9}}{R9} \cong \frac{V_T - V_{BE} - (V_T - 2\,V_{BE})}{R9} = \frac{V_{BE}}{R9}$$

Figure 7–103. Alternate circuit for electronic shutdown (National Semiconductor, *Linear Applications Handbook,* 1991, p. 319)

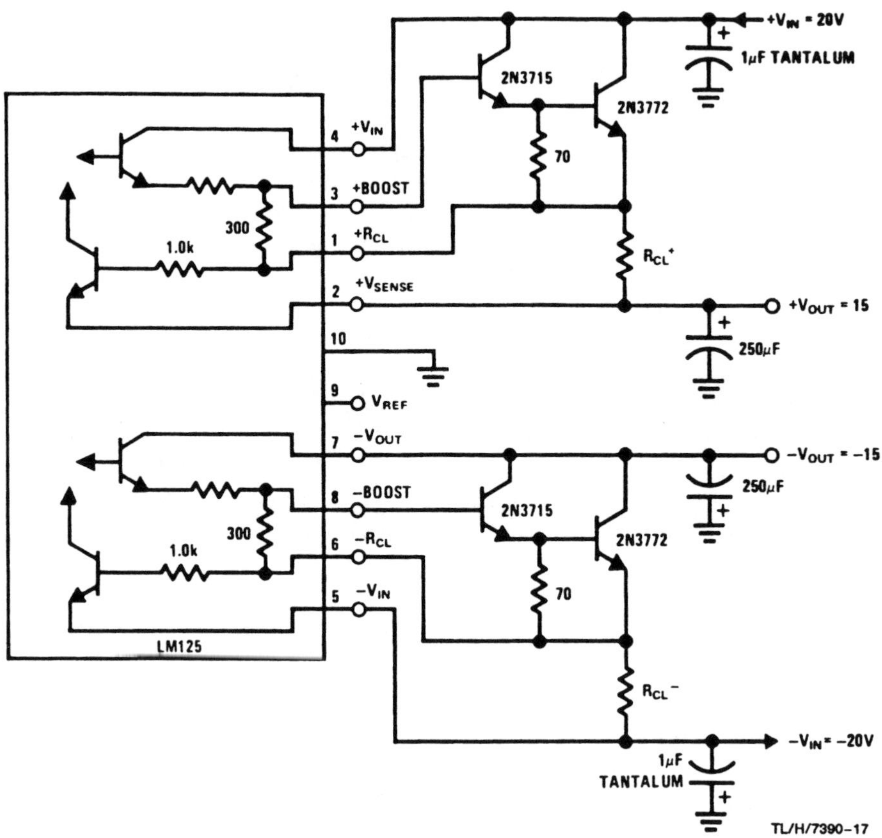

Figure 7–104. Boosted output using Darlington series-pass elements (National Semiconductor, *Linear Applications Handbook,* 1991, p. 311)

current can reach values to 10 A with very good stability. The external Darlington stage increases the minimum input-output voltage differential to 4.5 V.

When current-limit protection resistors are used as shown in Fig. 7–104 ($R_{CL}+$ and $R_{CL}-$), the maximum output current is limited by power dissipation of the 2N3772 (150 W at 25°C). During normal operation, this is $(V_{IN} - V_{OUT}) \times I_{OUT}$. However, under short-circuit conditions, the dissipation increases to V_{IN} times I_{SC}. The short-circuit output current is then:

$$I_{SC} = \frac{P_{MAX}\ (T_A = 25°C)}{V_{IN}} = \frac{150\ W}{20\ V\ (min.)} = 7.5\ A\ max.$$

Typical load regulation for the circuit is 40 mV from no load to full load. This assumes a junction temperature of 25°C and a pulsed load with 20 ms on and 250 ms off.

Notice that the load current could be increased to 10 A (or more) if the short-circuit current is less than the load current. However, this will probably require a foldback current-limiting circuit, such as described in Section 7.17.4.

7.17.7 Series-Pass Darlingtons with Current Foldback

Figure 7–105 shows the LM125 connected for a boosted output using Darlington series-pass elements (2N2890 and 2N3772), and with foldback current limiting. The design approach is similar to that discussed thus far in this section. However, in this design, the current contribution from the internal 300-ohm resistor is greater because of the $2V_{BE}$ drop across the Darlington pair.

Figure 7–106 shows the power dissipation in the external power transistor (Q_5, Q_7) for both positive and negative sides. Maximum power dissipation occurs between full load and short circuit, so the heat sink for the 2N3772 must be designed accordingly. The 2N3772 must be derated at 0.86 W/°C above 25°C. This corresponds to a thermal resistance junction-to-case of 1.17°C/W.

7.18 Simple Tracking Regulators

This section describes some simple tracking-regulator circuits based on the use of the classic National Semiconductor LM104/LM105. In systems with more than one regulated output voltage, it is often desirable to adjust all supplies with a single potentiometer. This results in savings by eliminating one or more potentiometer, as well as eliminating the need to adjust the supplies individually.

7.18.1 Tracking Positive Regulators

Figure 7–107 shows two LM105 regulators connected to provide +5-V and +15-V outputs. Both outputs track each other and are adjusted by a single potentiometer. Although the technique is not exact, the error is typically under 2%. As

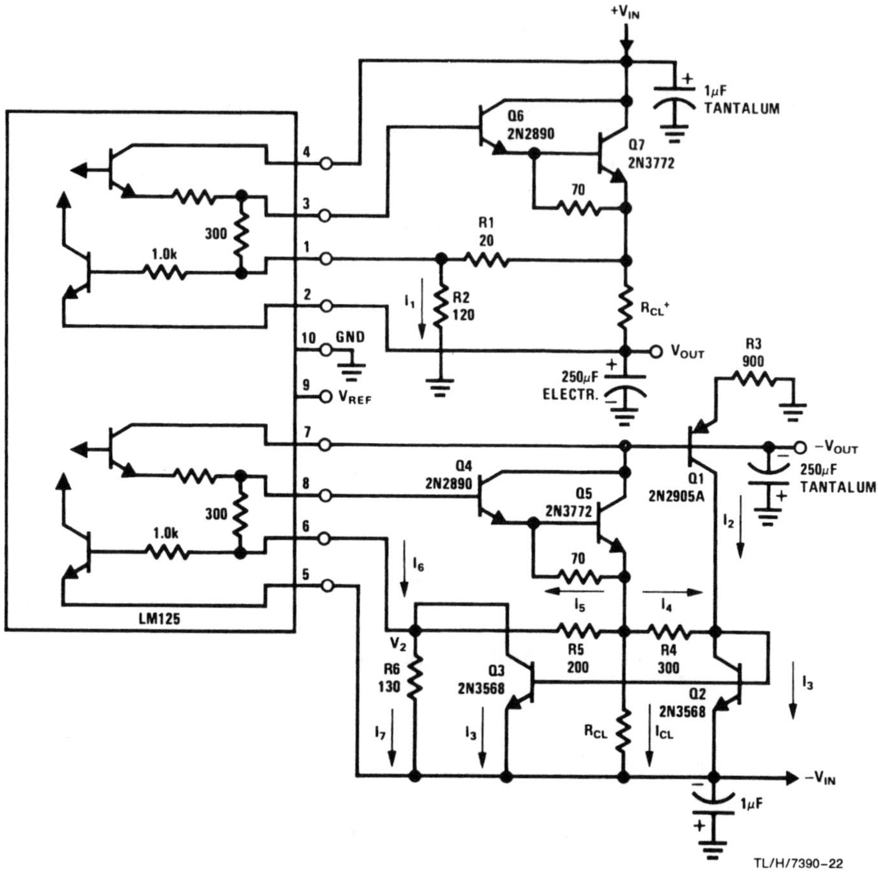

Figure 7–105. Boosted output using Darlington series-pass elements with current foldback (National Semiconductor, *Linear Applications Handbook,* 1991, p. 315)

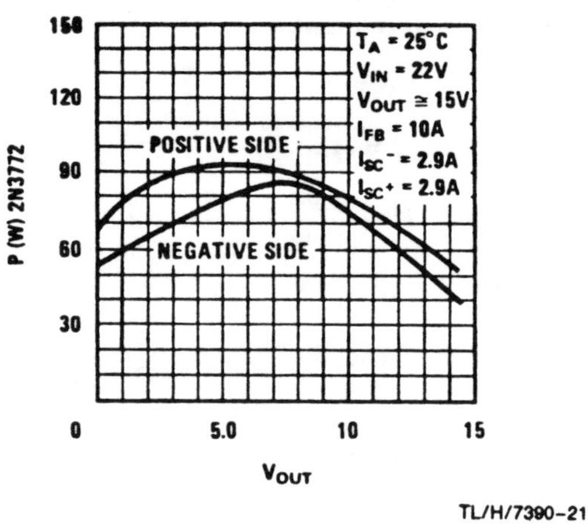

Figure 7–106. Power dissipation in external series-pass transistors Q_5/Q_6 (National Semiconductor, *Linear Applications Handbook,* 1991, p. 315)

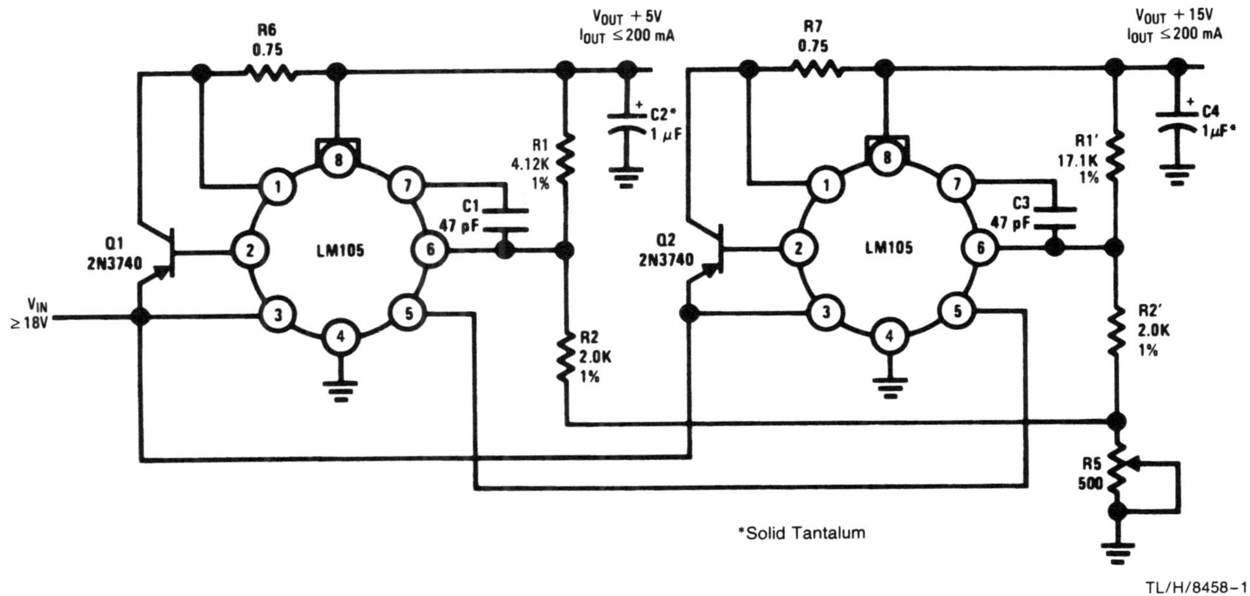

Figure 7–107. Tracking positive regulators (National Semiconductor, *Linear Applications Handbook,* 1991, p. 1191)

shown, the internal reference voltages for the LM105 regulators (available at pin 5) are tied together. This ensures that both regulators operate with the same reference voltage.

Output-divider resistors R_2 and R_2' are connected to ground through a common adjustment potentiometer R_5. Both regulators are adjusted by R_5 for any variation in the 1.6-V reference. Notice that the wiper of R_5 is connected to one side (ground end) of the potentiometer. If a rheostat connection were used, the wiper might open-circuit during adjustment, causing large transients in the output.

The calculations of values for the output divider are made with the consideration that the adjustment is not exact, and that two regulators are adjusted simultaneously. Resistors R_2 and R_2' are set at 2 k for convenience. R_1 and R_1' are then calculated for the desired output voltage using 1.6 V as the reference voltage.

To help compensate for the inaccuracies in adjustment, output voltages are calculated slightly off from the desired values. Resistor R_1 is calculated to give a 2% low output voltage on the 5-V regulator and a 2% high output voltage on the 15-V regulator, as follows:

$$R_1 = \frac{(V_{OUT} - 1.6\ V)\ 2000}{1.6\ V}$$

Using this equation, potentiometer R_5 adjusts both regulators (to within 2% of the desired output) for reference variations from 1.6 to 2 V. A 1.6-V reference produces outputs of 4.9 V and 15.3 V. If the reference is 2 V, R_5 is adjusted to 324 ohms and the output voltages are 5.1 V and 14.9 V. If the reference is near the typical value of 1.8 V, both outputs are within 1% of nominal.

The calculations do not account for resistor inaccuracies. If 1% resistors are used, there is an additional worst-case error of 2% for each regulator. Resistor errors are inherent in any type of tracking regulator system (even if the adjustment is theoretically exact!).

In theory, any number of regulators may be connected to a single adjustment potentiometer. The adjustment accuracy of this technique depends on the output-voltage difference among the regulators. The previous example was a severe difference, and has only 2% accuracy. With close output voltage, such as 12 V and 15 V, the error is much smaller.

As an example, if a 12-V regulator is calculated to ½% low, and a 15-V regulator is ½% high, with a 1.6-V reference, then both regulators are within ½% for reference variations from 1.6 to 2 V. This adjustment method is, of course, exact if two regulators have the same output.

7.18.2 Tracking Positive and Negative Regulators

Figure 7–108 shows LM104 and LM105 regulators connected to provide ±15-V outputs. Both outputs track, and no adjustment is required. In this circuit, the LM104 negative regulator tracks the LM105 positive regulator.

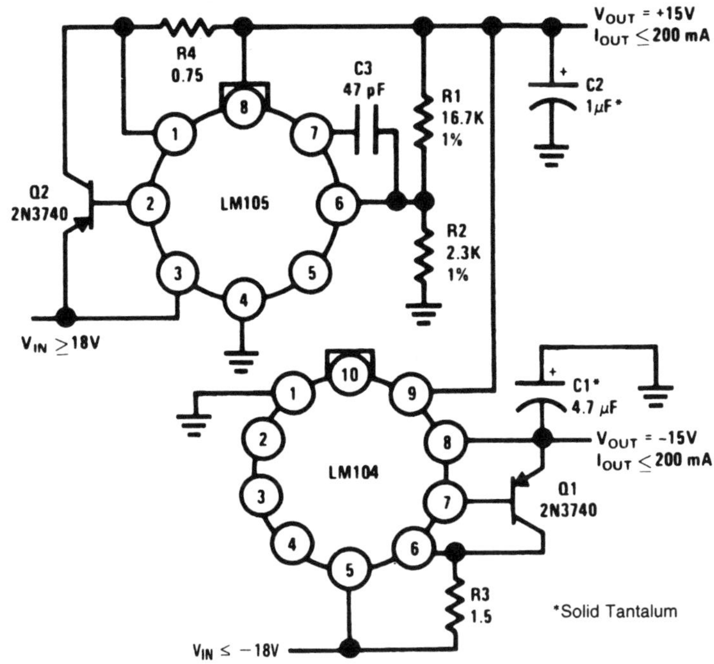

Figure 7–108. Tracking positive and negative regulators (National Semiconductor, *Linear Applications Handbook,* 1991, p. 1192)

The LM104 negative regulator is used as an inverting amplifier and provides several advantages over conventional op-amps. In normal operation, the LM104 is designed to drive boost transistors for higher output currents, as well as to provide current limiting. The frequency compensation used on the LM104 is optimized for transient response to line and load changes.

Operation of the positive/negative tracking is best understood by referring to the LM104 schematic in Fig. 7–109. The noninverting input of the internal amplifier (pin 1) is connected to ground. The positive 15-V reference is connected through an internal 15-k input resistor R_{16} to the inverting input. Feedback resistor R_{15} is also 15 k.

The R_{15}–R_{17} resistor network forms a unity-gain inverting amplifier, with a negative output voltage equal to the positive input voltage. The 15-k resistors in the LM104 are typically matched to 1%. This means that the output of both regulators may be set with 1% accuracy by changing R_1 in Fig. 7–108.

7.18.3 Tracking Regulators with Different Output Voltages

Figure 7–110 shows LM104 and LM105 regulators connected to provide +5-V and −15-V outputs. Again, the outputs track and no adjustment is required. In this

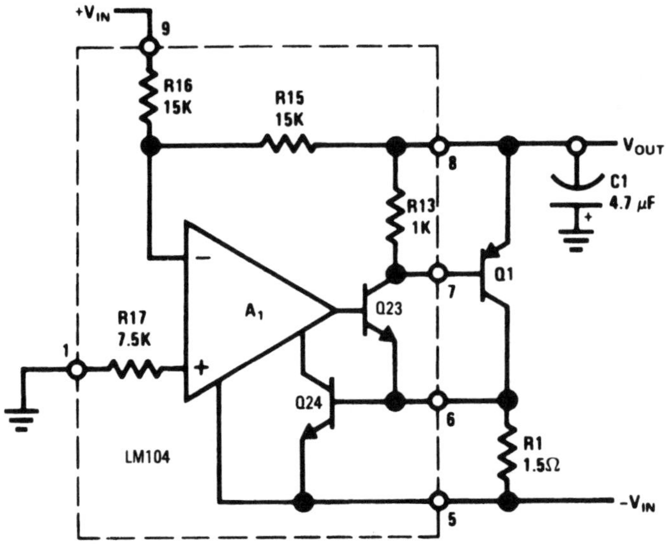

Figure 7–109. Functional schematic of LM104 used as an amplifier (National Semiconductor, *Linear Applications Handbook,* 1991, p. 1192)

circuit, the LM104 is used with inverting gain to produce negative voltages greater than the positive reference voltage. As shown, the noninverting input (pin 1) is not grounded, but tied to divider R_5/R_6 between the negative output and ground. The output voltage equals

$$V_{OUT} = V^{+}\left[\frac{R_6 + R_5}{R_6 - R_5}\right]$$

where V^{+} is the positive reference.

In this circuit, the line regulation and temperature drift are determined primarily by the positive reference. This reference must be a low-impedance source, such as the LM105 regulator, to ensure that current drawn by pin 9 of the LM104 does not affect the reference voltage.

Because the LM104 is connected to a positive voltage instead of ground, the regulator sees a total voltage equal to the sum of the unregulated negative input and the positive reference voltage. This reduces the maximum unregulated negative input voltage allowable, and should be considered during design.

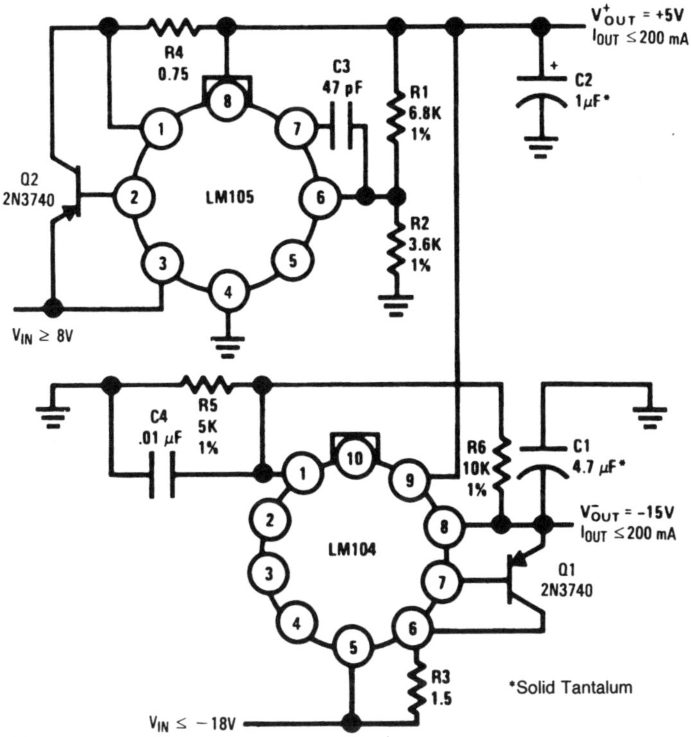

$$V^{-}_{OUT} = -V^{+}_{OUT}\left[\frac{R_5 + R_6}{R_6 - R_5}\right]$$

TL/H/8458–4

Figure 7–110. Tracking regulators with different output voltages (National Semiconductor, *Linear Applications Handbook,* 1991, p. 1193)

For Further Information

When applicable, the source for each circuit or table is included in the circuit or table title, so that the reader may contact the original source for further information. To this end, the complete mailing address for each source is given in this section. When writing, give complete information, including circuit title and description. Notice that all circuit diagrams and tables have been reproduced directly from the original source, without redrawing or resetting, by permission of the original publisher in each case.

Dallas Semiconductor
4401 S. Beltwood Pkwy.
Dallas, TX 75244–3292

EXAR Corporation
2222 Qume Dr.
PO Box 49007
San Jose, CA 95161–9007

GEC Plessey Semiconductors
Cheney Manor
Swindon, Wiltshire
United Kingdom 2N2 2QW

Harris Semiconductor
PO Box 883
Melbourne, FL 32902

Intel Corporation
3065 Bowers Ave.
Santa Clara, CA 95051

Linear Technology Corporation
1630 McCarthy Blvd.
Milpitas, CA 95035–7487

Maxim Integrated Products
120 San Gabriel Dr.
Sunnyvale, CA 94086

Motorola, Inc.
Semiconductor Products Sector
Public Relations Department
5102 N. 56th St.
Phoenix, AZ 85018

National Semiconductor Corporation
2900 Semiconductor Dr.
PO Box 58090
Santa Clara, CA 95052–8090

Optical Electronics, Inc.
PO Box 11140
Tucson, AZ 85734

Philips Semiconductors
811 E. Arques Ave.
PO Box 3409
Sunnyvale, CA 94088–3409

Raytheon Company
Semiconductor Division
350 Ellis St.
Mountain View, CA 94039–7016

Semtech Corpus Christi
121 International Blvd.
Corpus Christi, TX 78406

Siliconix Incorporated
2201 Laurelwood Road
Santa Clara, CA 95054

Texas Instruments
PO Box 1443
Houston, TX 77001

Unitrode Corporation
(Microsemi Corporation)
580 Pleasant St.
Watertown, MA 0217

Unitrode Corporation
8 Suburban Park Dr.
Billerica, MA 01821

Index